Essential Maths Skills for AS/A-level Business

Charlotte Wright and Mike Pickerden

Philip Allan, an imprint of Hodder Education, an Hachette UK company, Blenheim Court, George Street, Banbury, Oxfordshire OX16 5BH

Orders
Bookpoint Ltd, 130 Park Drive, Milton Park, Abingdon, Oxfordshire OX14 4SE
tel: 01235 827827
fax: 01235 400401
e-mail: education@bookpoint.co.uk

Lines are open 9.00 a.m.–5.00 p.m., Monday to Saturday, with a 24-hour message answering service. You can also order through the Hodder Education website: www.hoddereducation.co.uk

ISBN 978-1-4718-6347-9

First printed 2016
Impression number 5 4 3
Year 2020 2019 2018 2017

Typeset in India

Cover illustration: Barking Dog Art

Printed in India

Contents

Marketing

Operations

Human resources

Appendix

Introduction

Essential Maths Skills for AS/A-level Business will provide you with opportunities to practise the various quantitative skills you will need for your business exams. This includes statistical skills, such as correlation, together with using and interpreting graphs. At least 10% of the marks from your exam will come from level 2 (i.e. GCSE standard) quantitative skills.

This book assumes no prior knowledge of the mathematical concepts covered or of business concepts. Each section will introduce the topic and provide you with worked examples followed by guided and practice questions which build in difficulty.

You can use this book in various ways. You may want to look over pages 6-23 (Averages to Interpreting graphs) before starting your business course, to ensure you are familiar with the generic mathematical skills you will require. You can use this book throughout AS and A-level, alongside your textbook, for additional maths practice. In addition, you can use the questions as a resource to aid your revision when preparing for your exams. All questions are written in a business context, so answering the questions in this book will increase your understanding of business concepts as well as enhance your mathematical skills.

The publication is suitable for all exam boards. For information on which topics are relevant to your particular board and whether content appears in AS or A-level Year 2, see the cross-reference grid in the Appendix at the end of the book. A colour-coded system will help you determine which topics you need to study.

O	A green circle indicates that you need to be familiar with the entire contents of the section.
O	An orange circle signifies that some parts of the section are relevant and some are not – further details are given when this is the case.
O	A red circle means the section is not relevant to you.

To avoid losing easy marks, try the following approach:

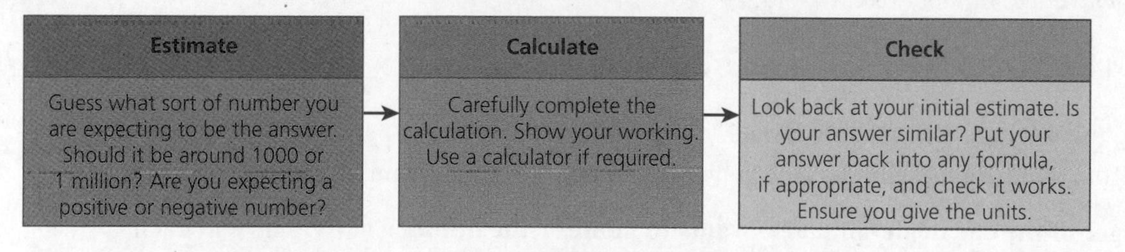

Some tips to remember:

- Unless otherwise stated, always round your final answers to two decimal places.
- When multiple calculations are required to reach the final answer, avoid rounding until you are at the final stage of the calculation.
- Always be sure to include units (e.g. £ or %) and show your working to get into good habits for the exam.

Full worked solutions to the guided and practice questions and exam-style questions can be found online at www.hoddereducation.co.uk/essentialmathsanswers.

1 Key mathematical skills

Averages

Averages are very useful in business. They enable managers to make comparisons and identify parts of the business that are performing well and parts of the business that are not. There are various types of average (median, mode, mean etc.). The main one you need to be able to calculate is a mean average. This average is calculated by adding up the numbers and dividing by how many numbers there are.

A Worked examples

a **A car dealership has five salespeople. The manager wants to review their performance. The table shows the number of cars each salesperson sold in July. Calculate the average number of cars sold by each salesperson in the sales team in July.**

Table 1.1

Name	Number of cars sold in July
Adam	15
Priya	30
Nick	50
Sam	25
Tom	35

Step 1: add the number of cars sold by all the salespeople to find the total number of cars sold by the dealership.

$15+30+50+25+35=155$ cars

Step 2: divide the total number of cars sold by the number of salespeople (in this case, five) to calculate the average number of cars sold.

$\frac{155}{5}=31$ cars

Calculating this average allows the manager to see which sales staff are performing above average (Nick and Tom) and who is selling a below average number of cars (Adam, Priya and Sam).

b **The manager of the car dealership also wants to monitor the number of days absent each employee had in July. Calculate the average number of days employees were absent in July.**

Table 1.2

Name	Number of days absent in July
Adam	6
Priya	0
Nick	0
Sam	4
Tom	0

Step 1: add all the days absent to find the total number of days absent:

$6+0+0+4+0=10$ days absent

Step 2: divide the total number of days absent by the number of salespeople (in this case, five) to calculate the average number of days absent.

$\frac{10}{5}=2$ days absent

B Guided questions

Copy out the workings and complete the answers on a separate piece of paper.

1 Below are last year's sales revenue figures for a chain of car dealerships.

Table 1.3

Store name	Value of sales in August (£)	Value of sales in September (£)
Bradford	75 000	80 000
Leeds	80 000	90 000
Huddersfield	105 000	100 000
Dewsbury	115 000	130 000

a Calculate the average value of sales in August.

Step 1: calculate the total value of sales in August.

$75000+80000+105000+115000=£375000$

Step 2: divide this number by the number of dealerships.

$\frac{£375000}{}=£$ __________

b Calculate the average value of sales in September.

Step 1: calculate the total value of sales in September.

$80\,000+90\,000+100\,000+130\,000=$ __________

Step 2: divide this number by the number of dealerships.

$\frac{\text{Total value of sales}}{\text{Number of dealerships}}=$ __________

c Identify the dealership(s) that recorded a below average value of sales in September.

Practice questions

2 An entrepreneur is doing an analysis of his competitors. He finds out the following information.

- Competitor A has 135 stores.
- Competitor B has 40 stores.
- Competitor C has 95 stores.

a What is the average number of stores owned by the entrepreneur's competitors?

b The entrepreneur has six stores. What is the average number of stores owned by all four businesses, including the entrepreneur's business?

3 A florist is investigating how long flowers last before they begin to droop and can no longer be sold to customers. He finds lilies last for 7 days, chrysanthemums last for 8 days, carnations last for 10 days and roses last for 5 days. He would like to collect information about tulips and daffodils next. Calculate the average time the florist's flowers last, based on the data he has available.

4 A manager and her team are looking at spending by departments. They work out that average spending in the current year has been £15 000. They looked at five departments:

- Marketing spent £20 000.
- Production spent £20 000.
- Human resources spent £5000.
- Customer services spent £10 000.

The sales team is the fifth department. Considering the average spend was £15 000, what must the sales team have spent?

Fractions

Fractions are useful in allowing businesses to see numbers in relation to other numbers. Common fractions include $\frac{1}{2}$ (a half), $\frac{1}{3}$ (a third) and $\frac{1}{4}$ (a quarter).

For example, if $\frac{1}{4}$ of the workforce are not happy at work, this means one out of every four employees is not happy at work. If $\frac{1}{3}$ of products made are shipped to America, this means one of every three products made are sold to the USA. To work out the size of a third, you would divide the number by three. To work out the size of a quarter, you would divide the number by four.

A Worked examples

a A supervisor of a clothes store works 40 hours per week. She complains to her line manager that a quarter of her time is taken up doing paperwork rather than serving customers or supporting her staff.

How many hours per week does she spend doing paperwork?

A quarter is $\frac{1}{4}$ so for every four hours she works, one of them is spent doing paperwork.

To calculate the number of hours spent on paperwork, divide the number of hours worked by 4:

$$\frac{40}{4} = 10 \text{ hours per week}$$

b Two thirds of employees ($\frac{2}{3}$) are full time. The rest are part-time employees. The business has 900 employees in total.

How many full-time employees does this business have?

Out of every three employees, two are full time.

It is easiest to work out one third of the number of employees first, by dividing the number of employees by 3.

$$\frac{900}{3} = 300 \text{ employees}$$

Two thirds would therefore equal $300 \times 2 = 600$. So 600 employees are full time.

B Guided questions

Copy out the workings and complete the answers on a separate piece of paper.

1 **Use the data in Table 1.4 to answer the questions.**

Table 1.4

Month	January	February	March	April
Number of units made	1 500	5 600	3 600	7 200
Proportion that needed to be scrapped due to poor quality	$\frac{1}{5}$	$\frac{2}{7}$	$\frac{1}{6}$	$\frac{2}{9}$

a **Calculate the number of units scrapped due to poor quality in January.**

One fifth is calculated by $1500 \div 5 = 300$ units

b **Calculate the number of units scrapped due to poor quality in February.**

Step 1: find one seventh: $5600 \div 7 = 800$ units

Step 2: multiply by two, to get two sevenths.

c **Calculate the number of units scrapped due to poor quality in March.**

Remember, to find one sixth, divide by six.

d **Calculate the number of units scrapped due to poor quality in April.**

Step 1: find one ninth: $7200 \div 9 = 800$ units

Step 2: multiply by two, to get two ninths.

2 **A business produces boxes of chocolates. Each box includes a mixture of four flavours. Each box contains 15 sweets in total. Use Table 1.5 to answer the questions.**

Table 1.5

Flavour	Caramel	Fudge	Orange	Strawberry
Desired proportion	$\frac{1}{3}$	$\frac{1}{5}$	$\frac{2}{5}$	

a **Calculate how many caramel sweets there should be in each box.**

One third is calculated by $15 \div 3 = 5$ sweets

b **Calculate how many fudge sweets there should be per box.**

To find one fifth, divide by five.

c **Calculate how many orange sweets there should be per box.**

Step 1: find one fifth of the box.

Step 2: multiply the answer to Step 1 by two.

d **Using the information you have calculated above, work out how many strawberry sweets there should be in each box.**

Step 1: work out how many caramel, orange and fudge sweets there should be per box, using your answers to the previous questions.

Step 2: now think about the number of sweets left that should be strawberry. Remember a box contains 15 sweets.

e **Express the number of strawberry sweets per box as a fraction.**

The fraction will be the number of strawberry sweets in a box divided by the total number of sweets in a box.

f **If a business produced 1000 boxes, how many caramel sweets would the business need to make?**

Step 1: look back to find how many caramel sweets were needed for one box.

Step 2: multiply your answer to Step 1 by the number of boxes you are being asked about.

C Practice questions

3 For a firm's factory to function at a satisfactory level, at least 40 staff must be present at any one time. A business has 252 staff employed and due to turn up for their shift that day, but due to a vomiting bug, only two ninths of the employees are in work.
   - a How many employees are in work that day?
   - b Can the factory function to a satisfactory level?
   - c The next day 63 employees arrive to work. Express this number as a fraction of the number of staff employed.

4 An employee works 16 hours per week and earns £7.50 per hour. She has been offered a promotion to team leader but she is not sure if she wants the extra responsibility. To encourage her to take the role, her manager has offered to increase her hourly wage by a third.
   - a Calculate what her new hourly wage rate would be.
   - b Calculate how much more she would earn per week at this wage rate, compared to her wage if she does not take the promotion.

5 A manager of a bakery has noticed it has a high wastage rate of flour. One day the manager measured that only four fifths of the flour was actually used to make products. The rest (3 kg) got spilled.
   - a How much flour was used to make products that day?
   - b After staff training, the manager reviewed the flour wastage again. The same amount was used but employees only wasted half as much flour as they did before. As a fraction, how much did they waste after the training?

Ratios

In a similar way to fractions, ratios show the relationship between two numbers. For example, a business might try to ensure that the ratio of men to women on its board of directors is at least 5 : 3. This would mean that for every five men it had on the board of directors, it would try to have at least three women. The same idea can be thought of as a fraction. If you look at the ratio, there are eight people or 'parts', i.e. 5 + 3. So another way to express the same idea would be to say that the board should be at least $\frac{3}{8}$ women.

Ratios are useful when a business wants to ensure it has the correct balance between two factors, e.g. the amount of cash it has and the amount of money it owes in the short term.

A Worked examples

a **A business has 1200 employees. 1000 are men and 200 are women.**

What is the ratio of male to female employees?

Step 1: this is 1000 : 200.

Step 2: this ratio could be made simpler by dividing both sides by a common factor, while still leaving the numbers as whole numbers. For example, you could divide both sides by 7, but that would give a ratio of 142.86 : 28.57. This is not very meaningful. It is very hard to visualise the workforce with this ratio.

Step 3: in this case both sides can be divided by the smaller number (200) to give a ratio in its simplest form of 5 : 1. It is now quite easy to picture this workforce. For every five men employed, there is only one female member of staff.

b **A business sells both men's and women's clothes, though it specialises in selling women's clothes. For every £2 revenue earned from selling men's clothes, the business makes £6 from selling women's clothes. The business earned £54 000 from the sale of women's clothes last year.**

i **Express the amount of money earned from women's clothes compared to men's clothes as a ratio.**

6 : 2 which can be made even simpler by dividing both sides by 2, giving a ratio of 3 : 1.

ii **How much revenue did the business make from selling men's clothes last year?**

Step 1: £54 000 was earned from selling women's clothes.

Step 2: the ratio of revenue from women's to men's clothing is 3 : 1.

Step 3: this means three times as much must have been earned from women's clothing than men's clothing.

Step 4: so revenue from men's clothing was £54 000 ÷ 3 = £18 000

B Guided question

Copy out the workings and complete the answers on a separate piece of paper.

1 **A gym offers both individual and corporate membership. Individuals can pay a monthly membership fee. Corporate membership is organised through businesses where firms can pay for a gym membership on behalf of their employees.**

The gym has 1000 members altogether. The ratio of individual to corporate members is 3 : 1.

a **Calculate the number of individual members the gym has.**

Step 1: there are four parts (3 + 1). One part equals 1000 ÷ 4 = 250 members.

Step 2: the number of individual members is therefore…

b **Corporate members pay £15 per month per person. Individual members must pay double this amount. What is the total amount of money received from membership each month?**

Step 1: find the revenue from corporate membership: 250 × 15 = £3750

Step 2: individual members pay $15 \times 2 = £30$ per month

Money earned from individual membership is £30 × ____________ = ____________

Step 3: add the revenue from corporate membership and individual membership.

C Practice questions

2 A business has 600 full-time employees and 40 part-time employees. What is the ratio of full-time employees to part-time employees, in its simplest form?

3 The owner of a manufacturing plant is analysing the amount the business spends on electricity and gas. She discovers the ratio of electricity to gas spending to be 2 : 1.

a The firm's gas bill was £2500. What was its electricity bill?

b What was the total amount spent on electricity and gas?

4 A newsagent sells a variety of products. He took £12 000 through the till last month. He makes a large amount of his money from selling magazines. The ratio of money earned from magazines to all other products is 3 : 2. How much money did he make from the sale of magazines last month?

5 A greengrocer has noticed the increase in popularity of organic food. He now sells equal amounts of organic and non-organic fruit and vegetables. Express this as a ratio.

6 A newspaper makes its money from the sale price of the paper and other businesses paying to advertise their products in the paper. A local newspaper tries to ensure that it has enough adverts in the paper to help the business be profitable. However, it worries that, if the paper has too many adverts, customers will be dissatisfied and turn to another paper. It aims for a ratio of content (articles etc.) to adverts of 7 : 2. The newspaper is 72 pages. A junior advertising executive says there are 8 pages of adverts. The rest is content.

a What is the newspaper's actual ratio of content to adverts?

b Is it on target?

c What might be the consequences of your answer to part **b**?

Percentages

Percentages are vital in business. In a similar way to fractions and ratios, they allow managers to compare numbers to other numbers. A percentage is a fraction that is always out of 100. If a business finds 10% of its employees are regularly late for work, this is the same as saying $\frac{10}{100}$ or one tenth are regularly late for work. If the firm had 200 employees, 20 are regularly late for work. Percentages are calculated by dividing the number you want to express as a percentage by the total number, then multiplying by 100.

Here is an example of how percentages allow managers to make comparisons and assess the actual scale of a problem. Imagine two businesses have eight employees leave as they felt bullied by their managers. No business wants any employees to leave for this reason, but with this information alone, it is hard to see the real magnitude of the problem. If you compare the number of leavers to the number of employees the businesses have in total, it is easier to make judgements.

Table 1.6

	Business A	Business B
Number of employees leaving as they feel bullied by managers	8	8
Total number of staff employed	16	800000

It is useful to work out the number of staff leaving due to bullying as a percentage of the total number of employees a business has.

This is done by dividing the number of employees who left due to bullying, by the total number of staff. You then multiply the number by 100 so it can be expressed as a percentage.

For Business A this is $\frac{8}{16} \times 100 = 50\%$

This is equivalent to half the staff leaving due to bullying. This is a real problem for Business A and should be investigated immediately.

For Business B the percentage is $\frac{8}{800000} \times 100 = 0.001\%$

The problem for Business B now seems nowhere as bad as Business A's situation.

A Worked examples

a Sarah completes her first business assessment. The maximum she could have scored was 40. Her mark was 35. What percentage did she score?

$$\frac{35}{40} \times 100 = 87.5\%$$

b A car dealership has five salespeople. Table 1.7 shows how many cars were sold in July by each member of the sales team.

Table 1.7

Name	Number of cars sold in July
Adam	15
Priya	30
Nick	50
Sam	25
Tom	35

Of all the cars sold, what percentage did Sam sell (to 2 decimal places)?

Step 1: the total number of cars sold is $15 + 30 + 50 + 25 + 35 = 155$ cars

Step 2: Sam sold 25 cars, so Sam's percentage is $\frac{25}{155} \times 100 = 16.13\%$

In some situations you might be given a percentage and want to work back from it. Imagine 60% of a shop's takings are from selling laptops. It makes £12 000 from selling laptops one month. How much money did it make in total? The information tells you that 60% of its takings is equivalent to £12 000. You have been asked to find 100%. One of the simplest ways of calculating this is by finding 1% initially, as shown here:

$60\% = £12000$

$1\% = £200$ (to find this divide both sides by 60)

$100\% = £20000$ (both sides were multiplied by 100 to get the full takings)

Always estimate, calculate and check. If you put these figures back into the percentage formula, you should get 60%:

$$\frac{£12000}{£20000} \times 100 = 60\%$$

A Worked examples

a Julie complains to her manager about workload. She works in a cake shop with two other employees, but Julie says she makes 70% of the cakes while the other two members of staff do very little and take very long breaks. Julie made 42 cakes last month. If Julie is correct, how many cakes did the other two members of staff make between them?

$70\% = 42$ cakes

$1\% = 0.6$ cakes (both sides were divided by 70)

$30\% = 18$ cakes (both sides were multiplied by 30 to calculate the remaining 30%)

b A confectioners uses 250 kg of fair trade sugar every week. The owner of the business says that 20% of the sugar it uses is fair trade but the rest is not fair trade. He would like to increase the amount of fair trade sugar to 80% in the future.

i How much sugar does the business use in total each week?

$20\% = 250$ kg

$1\% = 12.5$ kg (both sides were divided by 20)

$100\% = 1250$ kg (both sides were multiplied by 100 to calculate total sugar used)

ii If the owner reaches his goal, how much fair trade sugar would the business use each week?

1% of total sugar used $= 12.5$ kg

$80\% = 1000$ kg (both sides have been multiplied by 80)

B Guided questions

Copy out the workings and complete the answers on a separate piece of paper.

1 A business manufactures and sells bottles of shampoo and bottles of conditioner. It sold 500 bottles of shampoo and 300 bottles of conditioner in one week. Out of all the bottles sold, what percentage were bottles of conditioner?

The total number of units sold was $500 + 300 = 800$ bottles

2 It was Jason's job to check for faulty products before they were shipped to customers. One day 3% of the products he checked were faulty. He checked 2400 products in total. How many products were faulty? Complete the calculation:

Step 1: 100% = 2400

Step 2: 1% = __________

Step 3: 3% = __________

C Practice questions

3 A business has various members of staff working in different departments as summarised in Table 1.8. Calculate what percentage of the workforce is in the sales department.

Table 1.8

Department	Number of employees
Sales	2
Production	120
Personnel	3
Accounts	4
Health and safety	1

4 An employee works in a biscuit factory. They are told that between 20% and 22.5% of each biscuit's weight should be made up of chocolate. The employee analyses a sample of three biscuits. The results are shown in Table 1.9. Calculate which biscuits have an acceptable amount of chocolate on and which do not.

Table 1.9

Biscuit	Total weight of biscuit (g)	Weight of chocolate on biscuit (g)
A	16	3
B	17	3.7
C	16.5	3.8

5 A business has six stores. The managing director looked at the sales of each store as a percentage of sales of the whole business in one month. The results are shown in Table 1.10. Frome accounted for £26 000 worth of sales that month.

Table 1.10

Store name	Percentage of sales made by branch
Frome	13%
Bath	24%
Trowbridge	17%
Salisbury	5%
Bristol	20%
Swindon	

a What percentage of sales did Swindon account for?

b What was the total amount of money earned by the business that month?

c What was the total amount of money earned by Bristol?

6 A hotel has various costs ranging from the rent of the building, staff wages, laundry costs, utility bills and so on. A manager calculates that of all the costs, 20% are rent and 45% are wage costs. The business spends £2970 on wages per week.
   a Calculate the business's total costs per week.
   b Calculate what the business spends on rent per year.
   c If rent costs halved, then what would the firm's new total cost per week be?

Percentage change

Businesses and markets change. A key job of managers is to analyse this change and decide how to respond. Percentages are a useful tool here as they help managers see how large a change is. For example, imagine a business has an increase in its profits of £10 000 compared to the previous year. Is this a big change? Is this good news?

If you are a start-up company with £10 000 profit last year, this is an increase of 100% (profit has doubled)! If you are a large company with profits of £25m, this would represent an increase in profits of only 0.04%. Most businesses would see this as a concern, believing they should be growing much faster. This section will help you understand how to calculate a percentage change to allow you to see the relative size of a change. Percentage change figures can be positive, to show an increase in the size of a number, or negative, to show something is decreasing in size.

The percentage change formula:

$$\text{percentage change} = \frac{\text{change in the values}}{\text{original value}} \times 100$$

The change in the values is calculated by subtracting the original value from the new value. In some questions the change in the values may already be given to you, as in Worked example **a**. In some cases, you need to calculate the change in the values yourself, as in Worked example **b**.

A Worked examples

a **An employee is told he is getting a pay rise of £20 per week. He was earning £320. What is the employee's percentage increase in pay?**

The change is +£20. Divide this by the original pay and multiply by 100:

$$\text{percentage change} = \left(\frac{£20}{£320}\right) \times 100 = 6.25\% \text{ increase}$$

b **A business produced 2600 units last week and 3700 this week. What is the percentage increase in the number of units produced?**

$$\left(\frac{3700 - 2600}{2600}\right) \times 100 = 42.31\%$$

c **A salesperson sells 40 computers one week and 35 the next. What is the percentage change in the number of computers sold by the salesperson?**

$$\left(\frac{35 - 40}{40}\right) \times 100 = -12.5\%$$

Here, the answer is a negative number which indicates the number of computers sold has decreased.

In some cases, you might be told the percentage change figure and asked to work out the new value. There are various ways of calculating this. You might find it easiest to work through the method used in the previous section.

A business has sales of £150 one month. The next month, sales grew by 5%. Its new sales could be worked out as follows:

100% = £150

1% = £1.50

5% = £7.50

The new total sales = 150 + 7.50 = £157.50

Be sure to read exam questions carefully to see if the question asks for the new total or how much something has increased/decreased by.

Worked examples

a Zak thought he deserved a pay rise. When he started as a manager at the business he was in charge of 12 members of staff. He now manages 50% more people. How many people does Zak now manage?

100% = 12 members of staff

1% = 0.12 members of staff

50% = 6 members of staff

The total is the original number of staff Zak managed, plus the additional employees, i.e. 12 + 6 = 18 members of staff.

You may be aware that 50% is simply half of the total so, in this case, you may have been able to complete the problem in fewer stages.

b Humairah is concerned that her company's profit has fallen by 6% compared to the previous year. This year's profit is £160 458. What was the company's profit last year?

This question requires you to calculate the original profit.

We are trying to get back to the original profit figure before it fell by 6%.

We want to find 100% of the profit.

Current profit is 94% of the original figure.

94% = £160 458

1% = £1707

100% = £170 700

B Guided questions

Copy out the workings and complete the answers on a separate piece of paper.

1 **A manager is analysing how many complaints her stores received this year compared with last year. The results are shown in Table 1.11.**

Table 1.11

Store	Number of complaints last year	Number of complaints this year
St Helens	9	8
Bootle	15	12
Everton	20	14

a **Calculate the percentage change in the number of complaints received by the St Helens' store.**

$\left(\frac{8-9}{9}\right)\times 100 = 11.11\%$ decrease

b **Calculate the percentage change in the number of complaints received by the Bootle store. Complete this calculation. The first stage has been completed for you.**

Step 1: change in the values $= 12 - 15 = -3$

Step 2: use the formula

$$\text{percentage change} = \frac{\text{change in the values}}{\text{original value}} \times 100$$

with the value from Step 1 and the value from last year taken from Table 1.11.

c **Calculate the percentage change in the number of complaints received by the Everton store.**

Remember to find the change in the number of complaints over the year and divide by 20 (the original value).

2 **A toy store is looking at how its ranges have changed in popularity. The results are shown in Table 1.12.**

Table 1.12

Range	Number of units sold last year	Number of units sold this year	Percentage change in number of units sold compared to last year
Electronic games	2600	2800	7.69% increase
Bikes	550		12% increase
Musical instruments	430	400	
Wooden toys		420	5% increase

a **Calculate the number of bikes sold this year.**

Step 1: find 1% of 550.

Step 2: multiply the answer to Step 1 by 12 to find the increase.

Step 3: add the increase to 550 to find the number sold this year.

b What is the percentage change in the number of musical instruments sold?

- Note that the number of musical instruments sold has fallen so the percentage change figure will be a negative number.
- Use the formula to find the percentage change.

c How many wooden toys were sold last year?

Remember 420 toys = 105%

Step 1: find 1% by dividing 420 by 105.

Step 2: multiply your answer to Step 1 by 100 to get the number of wooden toys sold last year.

C Practice questions

3 An electricity provider noticed an increasing number of customers switching to their company. Customer numbers were 65 000 last year and are 68 000 this year. Calculate the percentage increase in the firm's number of customers.

4 A business hires out skips and must pay VAT. This is a tax paid to the government and is 20% of the price, at the time of writing. The business charges £90 for hiring a skip before VAT is added to the price.

a Calculate the price of hiring a skip after VAT.

b If the company hires out 30 skips, what is the total VAT it must pay?

5 A department incurs costs of £34 000 in one month. The department manager says this is unacceptable and the department must have spending of at least 35% less next month. The department spent £25 000 the following month.

a By what percentage has its spending fallen?

b What should its spending have fallen to in order to meet the target?

6 A business is having a sale. A junior sales assistant has been given information about the discounts as shown in Table 1.13 but must work out the new selling prices.

Table 1.13

Product	Price before sale	Discount
Jeans	£50	13% off
Skirts	£34	15% off
T-shirts	£12	12.5% off

a Calculate the new selling price for:

i jeans

ii skirts

iii t-shirts

b The sale is now ending and the employee must put prices back to their pre-sale price. However, he cannot remember their original prices. He only knows the discounted price and how much they were discounted by. Using the information in Table 1.14, calculate the original price of:

i coats

ii jumpers

iii suits

Table 1.14

Product	Price in the sale	Amount the product has been discounted by
Coats	£90	10%
Jumpers	£30	20%
Suits	£90	40%

7 A trade union is an organisation that tries to improve the rights of its members. Workers can choose whether they want to be in the trade union or not. The trade union representative for a company calculates that the number of members from the business who are a member of the trade union has changed in the last year. The results are shown in Table 1.15.

Table 1.15

	Total size of the workforce of the company	Percentage of the workforce who are in the trade union
Last year	1250	14%
This year	1300	15%

a Calculate the number of people in the trade union last year.
b Calculate the number of people in the trade union this year.
c Calculate the percentage change in union membership compared with last year.

Interpreting graphs

Data can be presented by businesses in various ways. Data may be in a table, as has already been shown in this book. Other options are to present data in charts/graphs such as a pie chart, line graph or bar chart. When interpreting data from charts, be careful to look at the axis and understand the units the data is presented in, e.g. pounds, percentage etc.

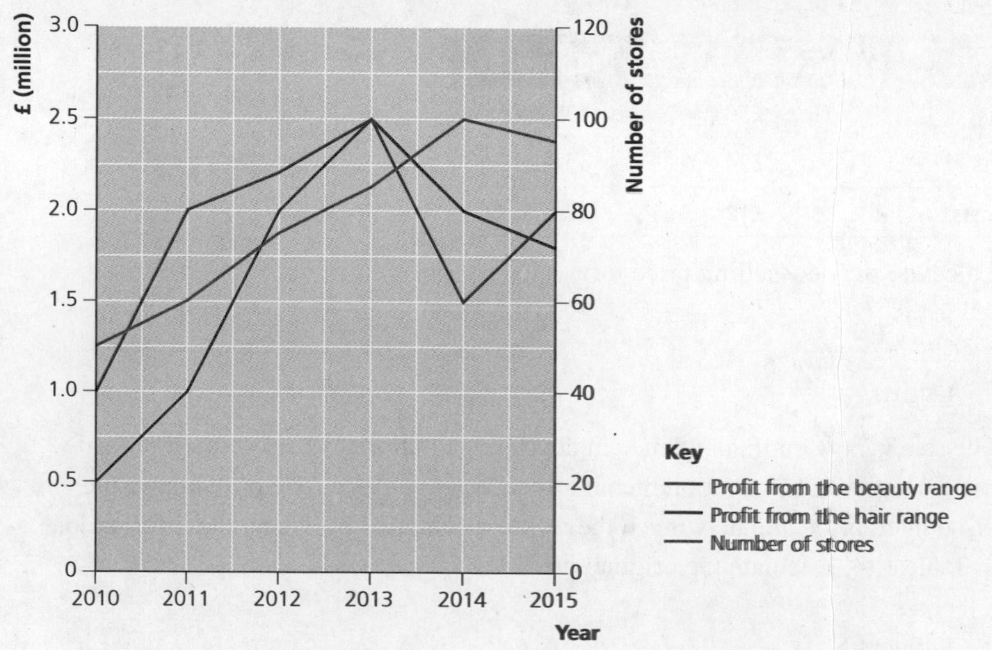

Figure 1.1 Data for Fabulous! Ltd

Figure 1.1 is a line graph showing some data for Fabulous! Ltd. A line graph is a common way of presenting data over time. The year is along the horizontal (x) axis. This graph has two vertical (y) axes. The number of stores should be read from the vertical axis on the right, e.g. in 2014 Fabulous! Ltd had 100 stores. The profit should be read from the vertical axis on the left.

A Worked examples

a Using Figure 1.1, calculate the total profit earned by Fabulous! Ltd in 2011.

Reading the data from 2011, it can be seen that the hair range raised £1m profit and the beauty range raised £2m, leading to a total of £3m profit.

Figure 1.2 is another line graph. However, there is an important difference between Figure 1.1 and Figure 1.2. Figure 1.2 shows the percentage change in price, i.e. the change in a variable, rather than its value.

Figure 1.2 shows the change in prices of Bspokz specialist bikes. A common student mistake is to say that the company lowered its prices in April and May. This is incorrect. As the graph shows the percentage change in prices, Figure 1.2 shows that prices have risen every month over the period shown. You can see this by the fact the line is always above zero. The downwards slope of the line after March represents prices rising but at a slower rate than previously.

b If Bspokz charged £1230 for a bike in April, what price did it charge in May? Use Figure 1.2 to help you answer.

Prices rose by 2% in May, compared to April.

1% of £1230 is calculated by $£1230 \div 100 = £12.30$

2% is $£12.30 \times 2 = £24.60$

The price charged in May was therefore $£24.60 + £1230 = £1254.60$

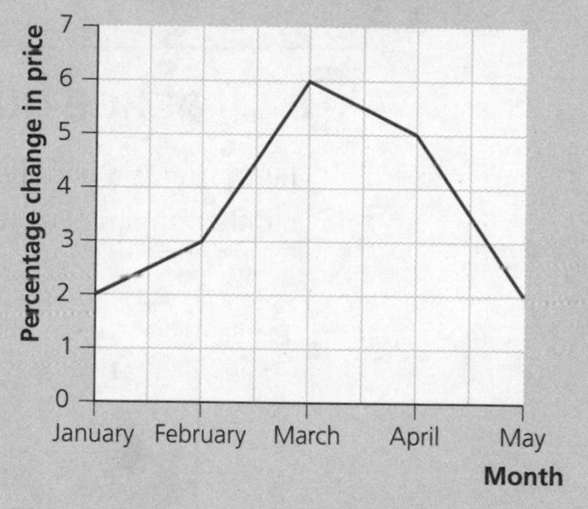

Figure 1.2 Percentage change in Bspokz prices compared to the previous month

Figure 1.3 is a pie chart. These are often used to show cross-sectional data – data from different subjects, e.g. different branches of a business, from one point in time. Pie charts are used to show numbers as a proportion of the total. Figure 1.3 shows the percentage of bus journeys provided by each firm, in a local area.

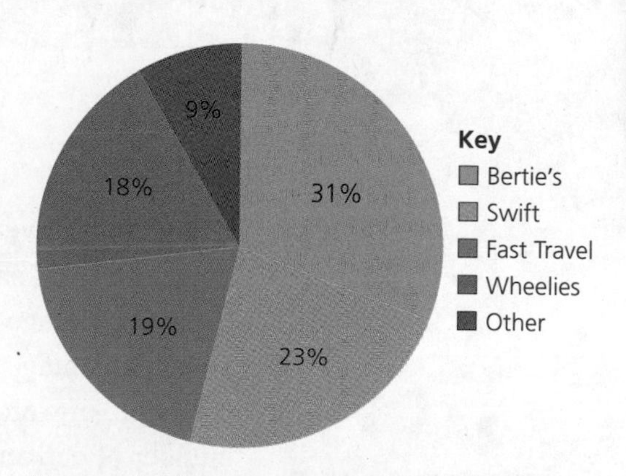

Figure 1.3 Percentage of bus journeys, by business

B Guided question

Copy out the workings and complete the answers on a separate piece of paper.

1 **Use Figure 1.3 to answer these questions.**

a **If Fast Travel provided 12 502 journeys to local people, how many journeys did Swift provide?**

Figure 1.3 shows that Fast Travel provided 19% of the journeys which is equal to 12 502 journeys.

Step 1: 1% of the journeys completed can be calculated by dividing 12 502 by 19, which gives 658 journeys.

Step 2: 23% of the journeys (Swift's per cent of the market) is therefore…

b **How many more journeys did Bertie's provide, compared to Swift?**

Step 1: calculate how many more journeys Bertie's provided compared to Swift in percentage terms.

i.e. 31 – 23 = ____________%

Step 2: from the previous question, you know 1% is equal to 658 journeys. Now multiply 658 by your answer to Step 1.

C Practice questions

2 Figure 1.4 is a bar chart which shows the number of Deliver-ease deliveries and the number of parcels which were delivered late, in recent months.

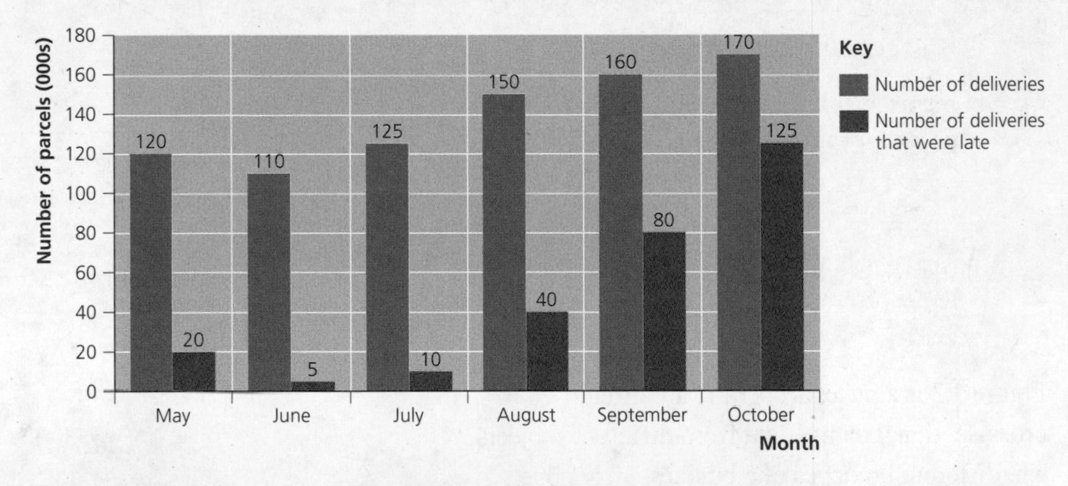

Figure 1.4 Deliver-ease delivery data

a Calculate the average number of parcels delivered by Deliver-ease per month, over the period shown. Note that the number of parcels is given in thousands.

b What percentage of deliveries in May were late?

c What percentage of deliveries in August were delivered on time?

d Describe, in words, the relationship between the number of deliveries and the number of deliveries that were late.

e What could explain your answer to part **d**?

3 Busy Clean sell cleaning services in the Worcester area. The marketing manager produced Figure 1.5 to help analyse performance.

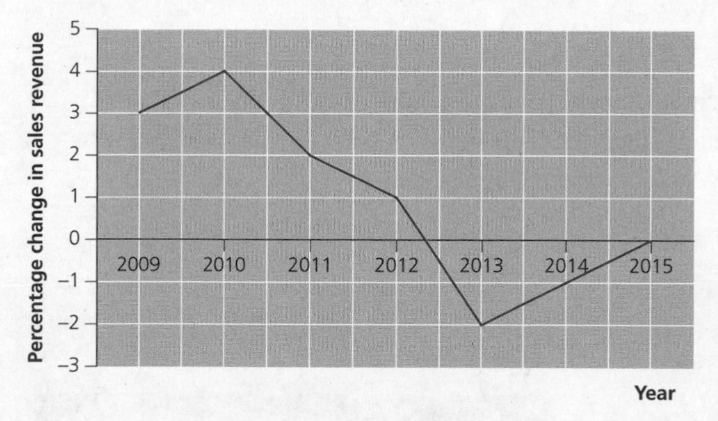

Figure 1.5 Sales data for Busy Clean

a In which years did sales revenue decrease?

b If busy clean raised £700 000 sales revenue in 2011, then what was sales revenue in 2012?

c In which year was sales revenue greatest?

Interpreting index numbers

Your business examination may require you to interpret index numbers. Ensure you are comfortable with how to calculate percentage changes before starting this section.

Index numbers can be a useful way of showing relative changes in data, over time. For example, imagine the number of units made in a factory one year was 54 654 654 800 and the next year it was 62 579 579 746. With large numbers like this, it is hard to see quickly how much the number has risen by. Using the percentage change formula you can calculate that output has increased by 14.5%, but very few people would be able to work that out in their head or have a grasp of the size of the change without getting their calculator out.

This is where index numbers become useful.

Table 1.16

	Factory output	Factory output index
Last year	54 654 654 800	100
This year	62 579 579 746	114.5

Table 1.16 shows the factory output in index form. Last year was used as the base year. This will be the year that all future changes are compared to. The index number of 100, on its own, means nothing. When the index number for this year's data is compared to last year's index number you can see that factory output has increased by 14.5%.

Whenever you calculate percentages with respect to 100, the process is very easy. You only need to subtract the numbers (114.5 – 100) rather than follow the full percentage change formula.

$$\text{percentage change} = \frac{\text{change in the values}}{\text{original value}} \times 100$$

$$\text{percentage change} = \left(\frac{114.5 - 100}{100}\right) \times 100 = 14.5$$

If you look at the second formula above, you can see dividing by 100 and then multiplying by 100 cancel each other out. This is why a base year is usually given a number of 100 as it is much easier to make comparisons to.

A Worked example

Use Table 1.17 to answer the questions.

Table 1.17

	Labour productivity index
2013	100
2014	105
2015	103

i **Explain what happened to labour productivity between 2013 and 2014.**

Labour productivity rose by 5%.

ii **Explain what happened to labour productivity between 2013 and 2015.**

Labour productivity rose by 3%.

iii **Explain what happened to labour productivity between 2014 and 2015.**

- With questions like this, you need to be careful. Labour productivity fell from 2014 to 2015. A common student mistake would be to say it fell by 2%.
- Remember, you can only subtract one number from the other to calculate the percentage change, when you are comparing to the base year (100). Otherwise you must follow the normal percentage change formula.

$$\text{percentage change} = \frac{103 - 105}{105} \times 100 = -1.9\%$$

This means that productivity fell by 1.9% from 2014 to 2015.

B Guided question

Copy out the workings and complete the answers on a separate piece of paper.

1 **The Consumer Prices Index (CPI) measures changes to average prices in an economy. Imagine Table 1.18 showed the Consumer Prices Index for two countries.**

Table 1.18

Year	Country A's CPI	Country B's CPI
2013	100	100
2014	102	103.5
2015	99	106

a **Calculate the percentage change in country B's CPI from 2013 to 2014.**
As your calculation is with respect to the base year (100) you only need to do one calculation.

b Calculate the percentage change in country B's CPI from 2014 to 2015.

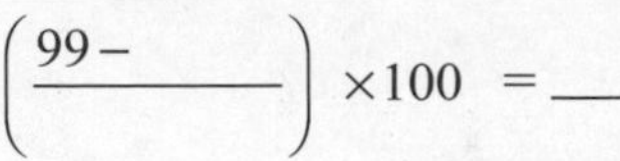

$$\left(\frac{106-103.5}{}\right)\times 100 = \_\_\_\_\_\_\_\_$$

c Calculate the percentage change in country A's CPI from 2014 to 2015.

$$\left(\frac{99-}{}\right)\times 100 = \_\_\_\_\_\_\_\_$$

Practice questions

2 A manager is studying a share price index, shown in Table 1.19.

a He says share prices grew by 15% from 2014 to 2015. Prove he is incorrect.

b How much did the share price index rise from 2013 to 2014, as a percentage?

c Explain what the share price index number of 100 for the base year of 2012 means.

Table 1.19

Year	Share price index
2012	100
2013	90
2014	110
2015	125

3 A commodity price index looks at the prices of various goods. Figure 1.6 shows the world prices of oil, steel and copper in index form. Time is shown on the horizontal (x) axis in quarters (every three months) so 2015 quarter 1 shows the average prices in January, February and March of 2015.

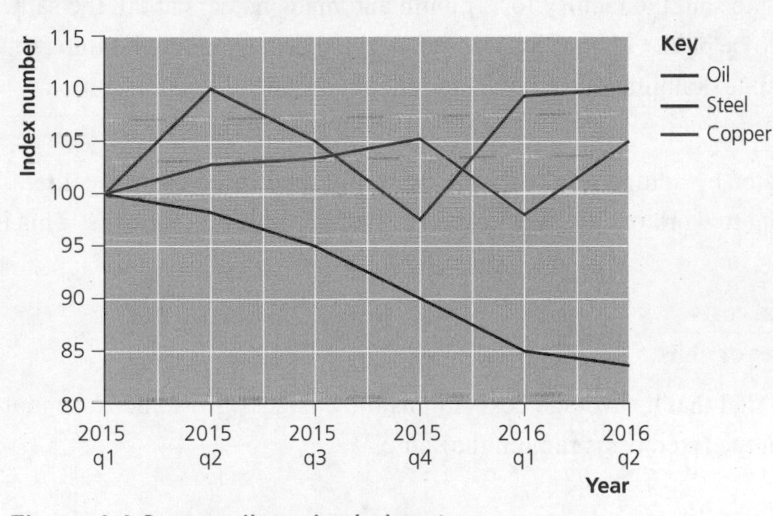

Figure 1.6 Commodity price index

a Explain the trend in steel prices.

b Calculate the percentage change in the price of copper from:

i 2015 quarter 1 to 2015 quarter 2

ii 2015 quarter 2 to 2015 quarter 3

iii 2015 quarter 2 to 2016 quarter 2

c Explain why it is not possible to tell, from this information alone, which commodity is the most expensive.

d Explain why Figure 1.6 might be useful to a business.

2 Finance

Costs

All businesses have costs. Costs are items of expenditure that the business must pay for, in order to make and sell its products. Examples include paying staff wages, buying raw materials and paying rent on the factory or store.

Costs can be classified in various ways. One way is to think of costs as fixed costs or variable costs. Variable costs are ones which increase as output increases. For example, the money spent on the wage bill for employees. If a car factory produced 1000 cars in a day and wanted to increase its output, it could hire new employees and perhaps double output to 2000 cars a day. However, these new employees would need to be paid, so the factory would see a rise in its variable costs. It would also see a rise in the cost of steel as it is now needing to buy twice as much.

Fixed costs are ones which do not vary with output. For example, if the car factory has the potential to manufacture a maximum of 5000 cars a day, the rent to be paid on the premises will not rise if output is increased from 1000 to 2000 cars (or decreased to 500 units either). The landlord will still expect the same amount of rent to be paid. If the factory owner decided to shut the factory for a month and made no cars at all, the same rent would still need to be paid. Other items classed as fixed costs include the line rental for the telephone. Despite doubling output, the amount paid to the telephone company would remain the same.

Total costs are calculated by adding total variable costs and total fixed costs together. A business may also want to work out average costs, i.e. cost per unit or unit costs. This is calculated by:

$$\text{unit cost} = \frac{\text{total costs}}{\text{number of units}}$$

A business will often find that its average costs fall as the business grows because more units are produced but the fixed costs remain the same.

A Worked examples

a **A business has £20 000 fixed costs per month and £3.55 variable costs per unit. The business produces 15 600 units in a month.**

i **What are the firm's total costs?**

Total variable costs are equal to the variable cost per unit multiplied by the number of units produced.

$$£3.55 \times 15600 = £55380$$

Total costs are total variable costs plus total fixed costs.

$$£55\,380 + £20\,000 = £75\,380$$

ii **What is the firm's average cost per unit?**

Here, total costs are divided by the number of units made.

$$\frac{75380}{15600} = £4.83$$

b A company has a total cost of £95 000 one month. Variable costs are £2.50 per unit and 26 500 units were made that month.

i **What was this company's total variable costs?**

£2.50 × 26 500 = £66 250

ii **What was this company's fixed costs?**

Subtracting total variable costs from the total cost gives the fixed cost.

£95 000 − £66 250 = £28 750

iii **Calculate the ratio of total variable costs to total fixed costs for this business, in its simplest form.**

£66 250 : £28 750

Both numbers are divisible by 1250 so in its simplest form, the ratio of total variable costs to total fixed costs is

£53 : £23

You might not know at first glance that both numbers are divisible by 1250, but from looking at the last few digits of each you might guess both are divisible by 250 which would give you a ratio of

£265 : £115

From this calculation, you can tell from the last digit that both numbers are divisible by five, which will give you the ratio in its simplest form.

Guided question

Copy out the workings and complete the answers on a separate piece of paper.

1 **A gym has £18 000 fixed costs per year. Variable costs per person, per visit are £0.25.**

a **If there are 1000 visits made to the gym in July, what are the total costs of running the gym that month?**

Step 1: calculate fixed costs per month (you have only been given the yearly figure):

Monthly fixed costs = £18 000 ÷ 12 = ___________

Step 2: calculate the total variable cost:

£0.25 × 1000 = ___________

Step 3: Add your total monthly fixed costs and total variable cost.

b **Calculate the average cost per visit for the gym, in July.**

Use your answer from part **a** and divide by the number of visits made to the gym in July.

C Practice questions

2 A business has £45 000 fixed costs per year. Variable costs are £4.23 per unit produced. Figure 2.1 is a chart showing the number of units manufactured over recent years.

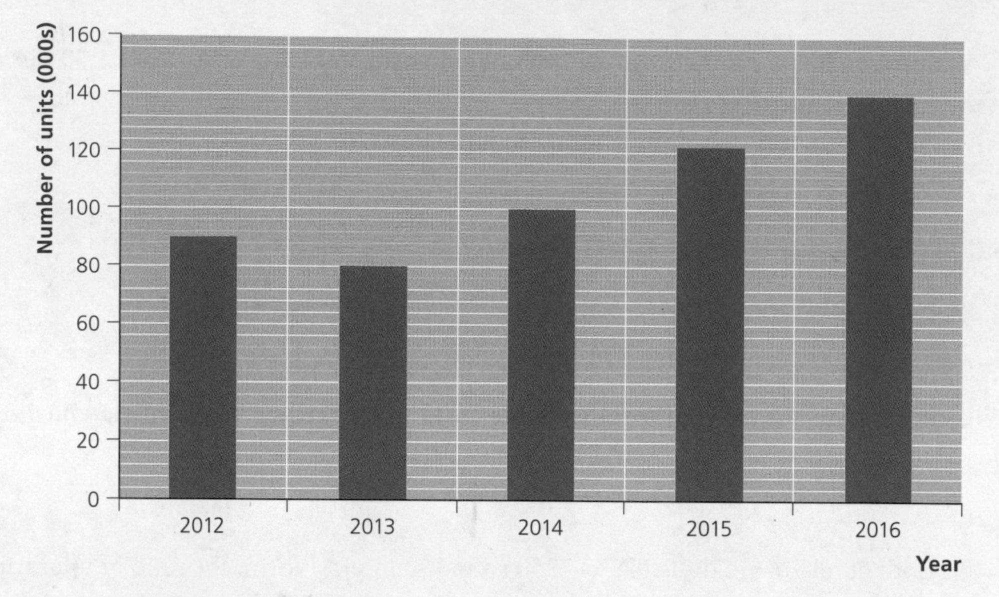

Figure 2.1 Output per year

a Calculate average costs in:

i 2014

ii 2015

iii 2016

b Calculate the percentage change in average costs from 2014 to 2016.

c What might explain the change in average costs from 2014 to 2016?

d Why is it unrealistic to assume fixed and variable costs per unit will remain the same from 2012 to 2016?

3 Last year a business had fixed costs of £57 000. This year, fixed costs have risen by 13%. Variable costs per unit have remained constant at £0.65 per unit made.

a Calculate average costs if 100 000 units were made last year.

b Calculate average costs if 100 000 units were made this year.

c Calculate the percentage change in unit costs this year, compared with last year.

Remember to use figures that have not been rounded to help calculate your answer.

4 A business has total costs of £265 000 this quarter (three months). Variable costs per unit are £6.70. 20 000 units were made.

a Calculate total variable cost for this business, this quarter.

b Calculate total fixed costs for this business, this quarter.

Revenue

Revenue (also known as turnover) refers to the money a business takes in from selling its products. It does not take into account the costs the business must pay out. Revenue is calculated using the following formula:

$$\text{revenue} = \text{price} \times \text{quantity sold}$$

A Worked example

A theme park has a busy Bank Holiday Monday selling 2000 adult tickets and 800 children's tickets. Adult tickets are £12 and children's tickets are £7. Calculate the theme park's revenue.

Step 1: revenue from adult tickets = £12 × 2000 = £24 000

Step 2: revenue from children's tickets = £7 × 800 = £5600

Step 3: total revenue = £24 000 + £5600 = £29 600

B Guided question

Copy out the workings and complete the answers on a separate piece of paper.

1 **Cheryl manages a group of swimming pools. She is monitoring the financial performance of three pools last month. Use Table 2.1 to answer the questions.**

Table 2.1

Branch	Price charged	Number of visits made by customers	Total revenue
Sheffield	£4.50	4 700	
Rotherham	£5		£20 800
Chapeltown		4 159	£24 954

a **What was Sheffield's total revenue last month?**

Look back at the revenue formula to help you.

£4.50 × ________ = ________

b **Calculate the number of visits made by customers, last month, in Rotherham.**

The revenue formula needs to be rearranged as follows:

$$\text{quantity sold} = \frac{\text{total revenue}}{\text{price}}$$

c **What price did Chapeltown charge, per customer, last month?**

The revenue formula needs to be rearranged.

$$\text{price} = \frac{\text{total revenue}}{\text{quantity sold}}$$

C Practice questions

2 A business sells caravans. Use the data in Table 2.2 to calculate its total sales revenue for last year.

Table 2.2

Product	Selling price	Number sold last year
Small caravan	£14 750	289
Large caravan	£18 000	460
Deluxe caravan	£22 000	156

3 A company sells 4000 units one year for £50 each. Total costs were £100 000. Calculate the company's total revenue that year.

4 Table 2.3 shows some data from a local newspaper. Though national newspapers are available, this business sells the only local newspaper in the area.

Table 2.3

Year	Price	Quantity sold
2014	£1	72 156
2015	£1.15	71 985
2016	£1.24	70 883

a In which year was total revenue largest?
b Calculate average revenue for 2014 to 2016.
c What trend do you notice in the data? What might explain this trend?
d What would you recommend the newspaper does to prices in future years?
e Can you tell from this data which year was most profitable?

Profit

Ensure that you are comfortable with the section on Costs (page 26) and the section on Revenue (page 28) before working through this section.

Profit is likely to be a word you associate strongly with businesses. Profit, in general, refers to the money a business can keep after it has paid for its costs and can be thought of as revenue minus total costs. However, there are various types of profit. By measuring profit in different ways, a business can pinpoint where it may be struggling financially.

The gross profit figure shows the amount of revenue remaining after direct costs of production, such as raw materials, are subtracted. These costs might be referred to as 'cost of sales' or 'cost of goods sold'.

gross profit = revenue − cost of sales

A Worked example

A business sells 5500 units one month. The price charged was £6.70 per unit. Cost of sales was £21 000. Calculate this firm's gross profit.

Step 1: total revenue = 5500 × £6.70 = £36 850

Step 2: gross profit = £36 850 − £21 000 = £15 850

Operating profit shows the amount of revenue remaining after both direct costs and other costs have been subtracted. These other costs are sometimes referred to as operating expenses or overheads. They refer to those costs which a business incurs from making and selling its products, but they are not directly attributable to production. Examples include rent on a store, utilities such as gas and electricity bills and marketing expenditure.

operating profit = gross profit − other operating expenses

A Worked examples

a **A business has gross profit of £15850. It has operating expenses of £12000. Calculate this firm's operating profit.**

Operating profit = £15 850 – £12000 = £3850

b **A company sells its products for £4.50. Last year it sold 70000 units. Cost of goods sold was £80000 last year and operating expenses were half that amount. Calculate the business's operating profit.**

Step 1: total revenue = £4.50 × 70000 = £315000

Step 2: total costs = £80000 + £40000 = £120000

Step 3: operating profit = £315000 – £120000 = £195000

Net profit is also known as profit for the year. 'Profit from other activities' refers to money raised from activities which are not the business's main way of generating income. For example, a chocolatier may sell off some of the stores it owns. If it makes profit from doing this, it will appear in the net profit figure but not in the operating profit figure as this money is not the chocolatier's normal way of making money. It would be misleading to place these earnings in the operating profit figure as it is unlikely money would be raised in this way year after year.

net profit = operating profit + profit from other activities – net finance costs – tax

'Net finance costs' refers to the money the business has spent on interest on loans and overdrafts. Tax is money given to the government, e.g. corporation tax which is a percentage of profit paid to the authorities.

A Worked example

A business has an operating profit of £3850. It has no profit from other activities. The business had £350 net finance costs. It must pay £770 in corporation tax.

Calculate this firm's profit for the year.

£3850 – £350 – £770 = £2730

B Guided questions

Copy out the workings and complete the answers on a separate piece of paper.

1 **Table 2.4 shows some financial data for a business. Use it to answer the questions.**

Table 2.4

Revenue	£2.9m
Cost of goods sold	£1.1m
Other operating expenses	£0.9m
Profit from other activities	£0.1m
Net finance cost	£0.04m
Tax	£0.18m

a **Calculate this business's gross profit.**

Gross profit = £2.9m – ____________ = £ ____________

b **Calculate the operating profit.**

Take your answer from part **a** and subtract the other operating expenses.

c **Calculate the business's net profit.**

Using your answer from part **b**:

Step 1: add on the profit from other activities.

Step 2: subtract the net finance cost and tax owed.

2 **Last year a business sold 10 000 units for £9 each. Cost of goods sold was £3 per unit. Other operating expenses were £78 000. Calculate the firm's operating profit for last year.**

Step 1: calculate total revenue: 10 000 × £9 = £90 000

Step 2: calculate cost of goods sold: 10 000 × £3 = ____________

Step 3: calculate gross profit: ____________ – ____________ = ____________

Step 4: operating profit: ____________ – ____________ = ____________

C Practice questions

3 Table 2.5 shows data for a firm's gross and operating profit for this year and last year.

Table 2.5

	Last year	This year
Gross profit	£1.46m	£1.78m
Operating profit	£0.9m	£1m

a Compared with last year, calculate the percentage change in:
- **i** gross profit
- **ii** operating profit

b What could have caused the change in gross profit?

c **i** Calculate the change in the firm's other operating expenses.
- **ii** What might have caused the change found in part **ci**?

4 Table 2.6 shows various pieces of financial information for a business this year. All data is given in thousands of pounds.

Table 2.6

Cost of goods sold	85
Other operating expenses	97.5
Profit from other activities	25
Finance costs	2
Tax	1

Last year the business generated £146 000 sales turnover. This year, revenue rose by 13%. For this year, calculate the firm's:

a gross profit

b operating profit

c net profit

5 A junior manager has expressed financial information for the past few years in index number form as shown in Table 2.7.

Table 2.7

Year	2013	2014	2015	2016
Revenue index	100	110	130	150
Cost of goods sold index	100	105	115	130
Other operating expenses index	100	102	105	110

a What was the percentage change in cost of goods sold between 2014 and 2016?

b The business had revenue of £70 000 in 2013, cost of goods sold was £25 000 and other operating expenses were £40 000. For 2013, calculate the firm's:

i gross profit

ii operating profit

c Using the information from the previous question, calculate for 2016, the firm's

i gross profit

ii operating profit

d The junior manager is concerned that cost of goods sold and other operating expenses are increasing. Give one reason why he should **not** be worried.

Profit margins

Ensure that you are comfortable with how to calculate a percentage and the material in the section on Profit (page 30) before studying profit margins. To help analyse profitability, it is often useful to compare a type of profit with the sales revenue it came from. Imagine two businesses have identical operating profit figures of £60 000. They appear, at first glance at Table 2.8, to be equally profitable. However, with extra information, a more insightful judgement can be made.

Table 2.8

	Company A	Company B
Operating profit	£60 000	£60 000
Sales revenue	£80 000	£1 000 000

If you express each company's operating profit as a percentage of its sales revenue, you can see which firm is more efficient at retaining a larger proportion of its sales revenue as operating profit.

For company A, its operating profit margin is $\frac{£60000}{£80000} \times 100 = 75\%$

For company B, its operating profit margin is $\frac{£60000}{£1000000} \times 100 = 6\%$

So although company B had the same operating profit as company A, its operating profit margin is far lower: 94% of its sales revenue is not kept as operating profit but is used to pay for fixed and variable costs.

Profit margins can be applied to the three types of profit discovered in the previous section. The formulae are as follows:

$$\text{gross profit margin} = \frac{\text{gross profit}}{\text{revenue}} \times 100$$

$$\text{operating profit margin} = \frac{\text{operating profit}}{\text{revenue}} \times 100$$

$$\text{net profit (profit for the year) margin} = \frac{\text{net profit}}{\text{revenue}} \times 100$$

Though this may look like a lot to remember, all three follow the same pattern of the profit you are looking at, divided by revenue (also known as turnover) multiplied by 100.

Businesses want their profit margin to be as high as possible, however what is judged to be a satisfactory profit margin will largely depend upon the industry. Some markets focus on having a large profit margin per unit sold, but may only sell a few units. Other markets focus on volume: selling large numbers of products but only having a small profit margin on each unit sold. To help make judgements about financial performance it is useful to know the firm's profit margins for previous years and the industry average profit margin or the performance of a close competitor. Without such information, it is very difficult to say whether a net profit margin of, say, 10%, is acceptable or not.

A Worked examples

a The data in Table 2.9 is taken from a company's income statement for this year. Using this, calculate the firm's gross, operating and net profit margins.

Table 2.9

Sales revenue	£5m
Gross profit	£3m
Operating profit	£1.5m
Net profit	£0.8m

$$\text{Gross profit margin} = \frac{3}{5} \times 100 = 60\%$$

$$\text{Operating profit margin} = \frac{1.5}{5} \times 100 = 30\%$$

$$\text{Net profit margin} = \frac{0.8}{5} \times 100 = 16\%$$

b An airline has a sales turnover of £4500m in one financial year. Its cost of goods sold was £2010m and its overheads were £1553m. Calculate the airline's gross and operating profit margins.

Step 1: calculate gross profit.

£4500m – £2010m = £2490m

Step 2: express this figure as a percentage of its sales revenue to give a gross profit figure.

$$\frac{£2490m}{£4500m} \times 100 = 55.33\%$$

Step 3: calculate operating profit.

£2490m − £1553m = £937m

Step 4: calculate operating profit margin.

$$\frac{£937m}{£4500m} \times 100 = 20.82\%$$

B Guided questions

Copy out the workings and complete the answers on a separate piece of paper.

1 A business sells 85 000 units for £2 each. Cost of goods sold is £60 000 and other operating expenses are £40 000. Calculate:

a the firm's gross profit margin

Step 1: calculate the firm's revenue: 85 000 × ______________ = ______________

Step 2: calculate gross profit by subtracting the cost of goods sold (£60 000) from the revenue.

Step 3: divide the gross profit by the revenue and multiply by 100.

b the firm's operating profit margin

Step 1: calculate operating profit by subtracting the £40 000 operating expenses from the gross profit.

Step 2: calculate the operating profit margin using the formula outlined at the start of the section.

2 A car company sells its vehicles for £14 500 each. It has a gross profit margin of 30%. Calculate the car company's cost of goods sold for each vehicle.

- The information tells you that 30% of every £1 of sales revenue is retained as gross profit.
- 70% is spent on direct costs such as the raw materials used in making the car.
- Therefore, to answer this question, you must find 70% of £14 500 to find the cost of goods sold per car.

C Practice questions

3 Hotel chain Flotel plc is comparing its recent financial performance with its closest competitor. Using the information in Table 2.10, calculate appropriate profit margins and assess which firm is most profitable.

Table 2.10

	Flotel plc	Ultimate Inns plc
Sales revenue	£3.4m	£2.9m
Gross profit	£1.9m	£1.9m
Operating profit	£1.3m	£1.25m
Profit for the year	£1.0m	£0.9m

4 A company's financial director calculates the profit margins for the last quarter shown in Table 2.11.

Table 2.11

Gross profit margin	35%
Operating profit margin	12.2%
Net profit margin	9%

That quarter, the business had £950 000 turnover. Using this information, calculate the company's:

a gross profit

b operating profit

c net profit

5 Use Table 2.12 to help you answer this question.

Table 2.12

	Last year	This year
Average selling price	£338	£340
Number of cakes sold	105	116
Cost of goods sold	£4 200	£5 000
Other operating expenses	£31 000	£33 000

Compared to last year, calculate the change in this wedding cake store's:

a gross profit margin

b operating profit margins

6 A business generated £37 789 operating profit last year with an operating profit margin of 23%. Calculate this firm's sales revenue last year.

7 Trendsetter Ltd has sales revenue of £90 000 and cost of goods sold of £50 000. Compare this company's gross profit margin to the industry average. The other firms in this industry's gross profit margin figures are listed in Table 2.13. Include Trendsetter Ltd's gross profit margin in the industry average figure.

Table 2.13

Business A	40%
Business B	33.7%
Business C	51%
Business D	23%
Business E	44%

Contribution

The last two sections have focused on studying profit which is an important business objective. However, for many smaller firms, their main aim might not be to make lots of profit, but to break-even. A firm's break-even point refers to the number of units a business must sell in order to make neither a profit nor a loss, i.e. to take just enough revenue to cover all costs.

How is the break-even point calculated? This is discussed in the next section. However, to be able to calculate break-even, we first need to look at 'contribution'.

Example

Imagine a hotel has one guest. They pay £50 for a night's stay in the hotel. Variable costs are £10 per guest per night. Fixed costs per day are £100.

Think about where the £50 (the hotel's revenue) goes. £10 of it is used to pay the variable costs the hotel incurred due to the fact the guest was staying, e.g. washing the guest's sheets. The remaining £40 can go towards paying the fixed costs of the hotel, e.g. the rent. This guest contributes £40 to the hotel's fixed costs. If the hotel had two guests, total contribution would be £80 (£40 × 2).

If the hotel had three guests, the total contribution would be £120. This is higher than the £100 fixed costs the hotel had so the extra £20 is profit.

Here are some useful formulae, illustrated by the above example:

- contribution per unit = price – variable cost per unit
- total contribution = contribution per unit × number of units sold *or*
- total contribution = revenue – total variable costs
- profit = total contribution – fixed costs

So, contribution refers to the amount of revenue a business has to put towards its fixed costs, after it has paid its variable costs.

A Worked examples

a A supermarket sells tins of beans for 50p each. Variable costs are 40p per can. One month the supermarket sells 2500 tins of beans.

i Calculate the contribution per unit.

Subtract variable costs per unit from the selling price:

50p – 40p = 10p

ii Calculate the total contribution.

Multiply contribution per unit by the number of units sold.

£0.10 × 2500 = £250

b A business has sales revenue of £2000 in one month. Its variable costs, in the same time period, were £700 and fixed costs were £1000.

i Calculate the total contribution.

Variable costs are subtracted from the sales revenue:

£2000 – £700 = £1300

ii Calculate the business's total profit.

Fixed costs are now subtracted from total contribution to see how much money the business has left over after paying its fixed costs:

£1300 – £1000 = £300 profit

B Guided questions

Copy out the workings and complete the answers on a separate piece of paper.

1 A bar sells pints of beer for £2.90. Its variable costs are £1.35 per pint. In a year the pub sells 30 000 pints and has fixed costs of £41 000.

a Calculate the contribution per unit.

For one pint of beer this is £2.90 – ______________ = ______________

b Calculate total contribution.

Multiply the contribution of one pint of beer by the total number of pints sold.

c What is the pub's profit?

Subtract the fixed costs from the total contribution to see how much money is remaining that the pub can keep as profit.

2 A business sells 20 000 units. Its unit contribution is £1.25 and selling price is £2. The total profit made is £5000.

a Calculate the firm's total variable cost.

Step 1: the difference between the selling price and the unit contribution must be the variable cost per unit, i.e. £0.75.

Step 2: now calculate total variable cost, considering that 20 000 units were sold.

b Calculate the size of the business's fixed costs.

Step 1: calculate total contribution:

£1.25 × ______________ = ______________

Step 2: total contribution must have been £5000 greater than fixed costs as the business had £5000 profit after all costs had been paid. Using this information, try to calculate its fixed costs.

C Practice questions

3 An ice cream van has fixed running costs of £100 per day. Variable costs are £0.50 per ice cream sold. One sunny Saturday the van sells 574 ice creams for £1.40 each.

- **a** Calculate contribution per ice cream.
- **b** Calculate total contribution.
- **c** The next day is not as sunny and only $\frac{2}{7}$ as many ice creams are sold. Calculate the new number of ice creams sold.
- **d** Calculate the new total contribution figure.
- **e** Calculate the change in profit from Saturday to Sunday.

4 A business has a pricing policy of selling its products for its variable cost plus 30%. It sells 3500 units. Its variable costs are £1.70 per unit.

- **a** Calculate contribution per unit.
- **b** Calculate total contribution.

5 Fixed costs are £80 000 for a company, per year. Contribution per unit is £2.78 and 25 000 units are sold. Calculate this company's profit.

6 A golf course has fixed costs of £14 000 per month. With 1000 members paying £62.50 per month to be a member, the course made £21 000 profit last month. Calculate the contribution per member.

Break-even

Break-even refers to the number of units which must be sold in order for the firm's profit to be zero, i.e. where total revenue and total costs are equal.

It is very useful for businesses to have an idea of how many units must be sold in order to break even. It can help managers decide whether a venture is viable. For example, if an entrepreneur wanting to set up a café knew he would have to pay £7000 rent per week, could charge £2 for a cup of coffee and had £1.50 variable costs per cup of coffee, he could work out that he would need to sell 14 000 cups per week just to cover his costs. If he opened between 9 a.m. and 5 p.m. every day, this is the equivalent to 250 cups sold per hour. Knowing this might help him decide whether this is realistic or not.

The break-even point tells a business how many units it must sell in order to break even. It is calculated using the following formula:

$$\text{break-even output} = \frac{\text{fixed costs}}{\text{contribution per unit}}$$

Remember, contribution per unit is calculated using this formula:

contribution per unit = selling price – variable cost per unit

Many businesses are interested in their break-even point, but are not content with simply breaking even. They are hoping to sell additional units so they make a profit. They want a margin of safety.

A margin of safety refers to the additional number of units a business sells, over and above their break-even point. The formula is:

margin of safety = planned output – break-even output

Worked example

A business has fixed costs of £100 000 per year. It sells its products for £4 per unit and has variable costs of £1.50 per unit. One year the business sells 60 000 units.

i Calculate this firm's break-even point.

$$\frac{£100000}{£4-£1.50} = 40000 \text{ units}$$

ii Calculate this firm's margin of safety.

60 000 – 40 000 = 20 000 units

This means the business can keep all the revenue it takes on the final 20 000 units it sells.

These ideas can also be expressed on a chart.

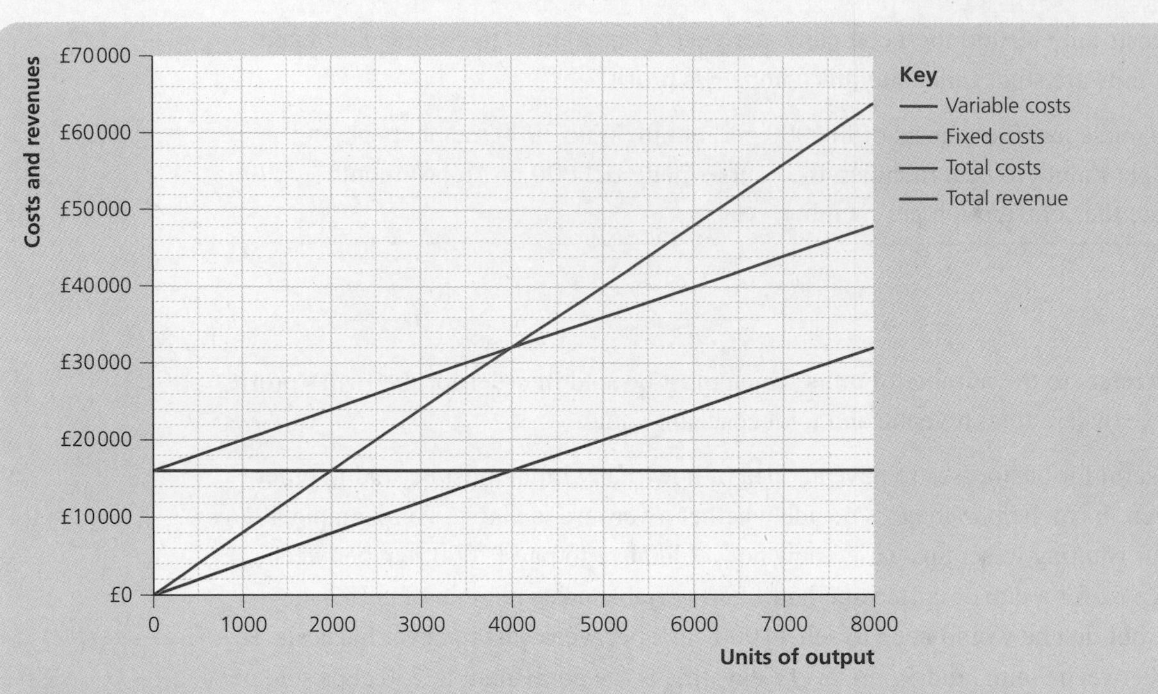

Figure 2.2 Break-even chart example

In Figure 2.2, the fixed costs line is horizontal as it does not change as output changes. As variable costs increase as output rises, the variable costs and total costs lines slope upwards. The total revenue line also slopes upwards as the more a business sells, the more revenue it will have.

iii Using Figure 2.2, calculate the firm's break-even point.

Step 1: to find the break-even point, look at the total costs and total revenue lines.

Where they are equal (cross) this indicates the number of units that must be sold in order to break even.

Step 2: in this instance 4000 units must be sold to break even.

iv Calculate the size of the firm's profit or loss if 6000 units are sold.

Step 1: read off the total costs at 6000 units: £40 000

Step 2: now read off total revenue at 6000 units: £48 000

profit = total revenue – total costs

So profit at 6000 units is £8000.

v Show how the break-even point changes if the business increases its price by £4 per unit.

Step 1: calculate the current price, using the total revenue line. Pick an output point to help you do this, e.g. 1000 units.

If 1000 units are sold, revenue would be £8000. This indicates that revenue per unit (i.e. the selling price) is £8000 ÷ 1000 = £8

Step 2: if price was to rise by £4, this would give a new selling price of £12. The total revenue line would become steeper. To help you plot this, all you need is two points. One point is easy: at zero output, total revenue will still be zero. Now pick another point and calculate total revenue, to help you draw a straight line. For example, at 5000 units, current revenue is £40 000. Now at the new selling price, the revenue would be 5000 × £12 = £60 000.

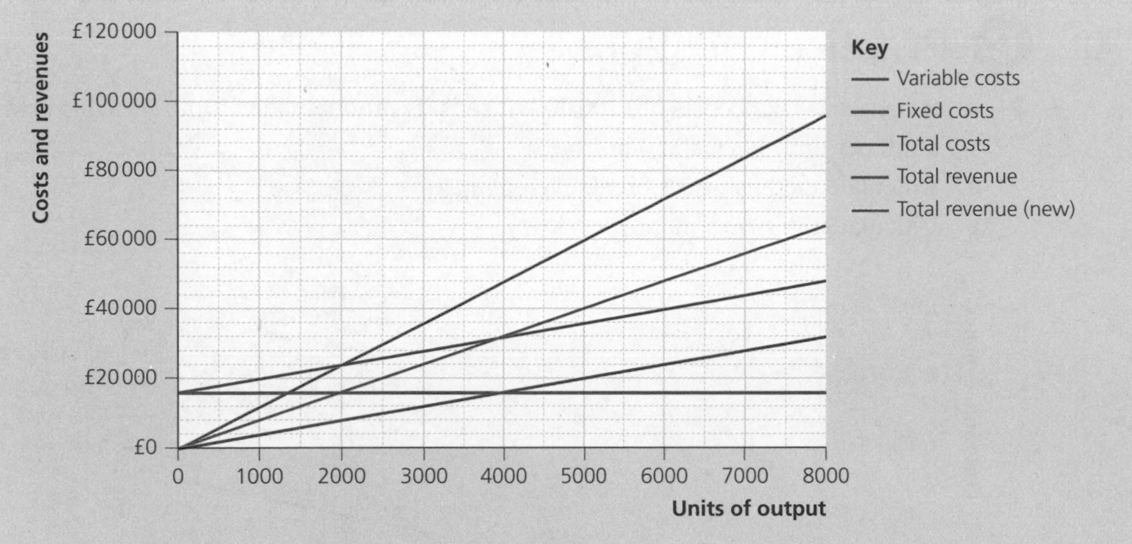

Figure 2.3 Break-even chart example amended

As shown on the amended break-even diagram, the new total revenue line has become steeper and the break-even point has fallen to 2000 units. So a £4 price rise halves the break-even output required for this business.

B Guided questions

Copy out the workings and complete the answers on a separate piece of paper.

1 The owner of a pottery business is trying to calculate her break-even point.

Table 2.14

Number of units sold	7100
Fixed costs	£34000
Selling price	£13
Variable costs per unit	£8

a Calculate the firm's break-even point.

Step 1: contribution per unit = £13 – £8 = ______

Step 2: break-even output = $\frac{\text{\_\_\_\_\_\_\_\_\_\_}}{\text{contribution per unit}}$

b Calculate the firm's margin of safety.

margin of safety = 7100 – break-even output

2 A business has a break-even point of 5000 units. It sells 6000 units. It sells its products for £6 per unit. Calculate the firm's profit.

At 5000 units, total revenue is equal to total costs. On the additional 1000 units sold, all revenue earned can be retained as profit.

C Practice questions

3 A business has fixed costs of £100 000. It sells its products for £1000. Variable costs are £500 per unit. It sells 250 units.

a Calculate the business's break-even point.

b Calculate the firm's margin of safety.

4

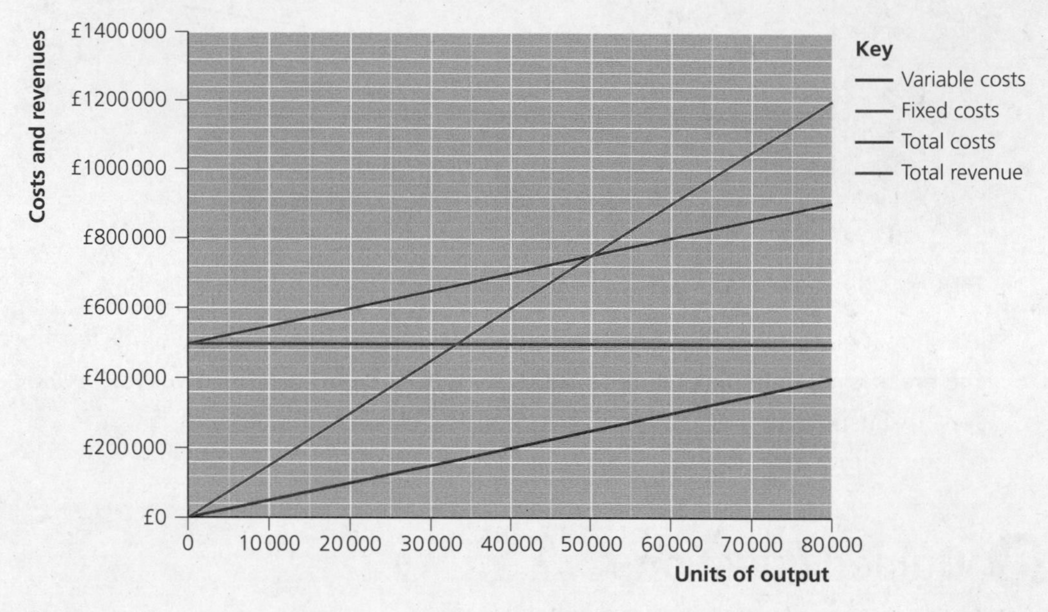

Figure 2.4

Using Figure 2.4, calculate:

a the profit/loss made at 30 000 units

b the profit/loss made at 70 000 units

c the business's break-even point

d the margin of safety if 50 000 units are sold

e how the break-even point changes if fixed costs rise by £100 000

5 The following data is from a home removals business:

Fixed costs	£20 000
Price	£750
Variable cost per home removal	£250

a If fixed costs rose by 20%, how would the number of removals the firm needs to do in order to break even change?

b Compared to the original situation, calculate how the break-even point would change for this business if it increased prices by one third.

c Compared to the original situation, calculate the change in break-even point if variable costs halved.

6

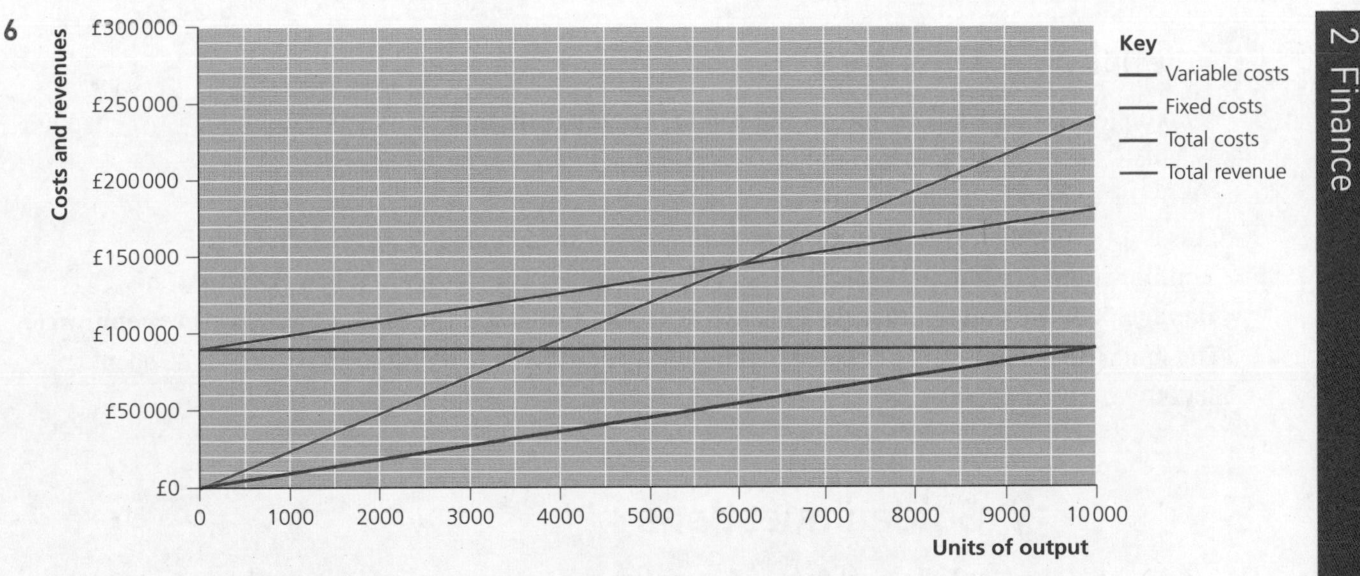

Figure 2.5

Use Figure 2.5 to calculate:

a the range of output where the business would make a loss
b the margin of safety if 10 000 units are made
c total profit if 10 000 units are made
d total contribution if 10 000 units are made

7 A new entrepreneur is having financial difficulties. To help resolve this, she decides to calculate her break-even point. She has £15 000 fixed costs. She sells her product for £10 each. Her variable costs are £12 per unit. Calculate her break-even point and explain why she is struggling financially.

Budgets and variance analysis

In business, a budget refers to a numerical target. A manager might have an expenditure budget, e.g. do not spend more than £20 000 this quarter. They might have an income budget, e.g. generate £100 000 worth of sales revenue this quarter. From this, often a profit budget can be inferred which is calculated by the income budget less the expenditure budget. In the above example, the manager has an £80 000 profit budget.

Managers often do not exactly meet their targets. They often have a variance: a difference between their budget and the actual figure. A manager might have a favourable variance if he or she underspends on their expenditure budget, having a lower actual figure than budgeted figure or if they exceed their income budget.

A manager might have an adverse variance if they have a higher actual expenditure compared to their expenditure budget or a smaller actual income than their income budget.

In summary, if the difference between the budget and the actual figure means profits will be higher for the business, this is a favourable variance. If the difference between the budget and actual figure means profits will be lower than planned, this is known as an adverse variance.

A Worked examples

a **A manager has an income budget of £90 000. Their actual income is £98 000.**
Calculate the manager's variance. State whether this is adverse or favourable.
£98 000 − £90 000 = £8000 variance
This is a favourable variance as the business has more income than it had been targeted.

b **A junior manager had an expenditure budget of £8000 one quarter. When reviewed, his manager said he had an adverse variance of £1500. Calculate the junior manager's expenditure.**
The junior manager must have spent his entire expenditure budget plus the variance so he spent £8000 + £1500 = £9500.

B Guided questions

Copy out the workings and complete the answers on a separate piece of paper.

1 **A business has a policy of using historical budgeting. This year, each department gets the same budget as the previous year plus inflation (the rate at which the general price of goods and services is increasing). Imagine inflation is 2.2%. Table 2.15 shows last year's budget.**

Table 2.15

Department	Last year's budget
Shoes	£36 000
Handbags	£31 000
Clothes	£98 760

a **Calculate the new budget for the shoe department.**

Step 1: 1% of the shoe department's budget is £36000 ÷ 100 = £360

Step 2: 102.2% of the shoe department's budget is £360 × 102.2 = £36 792

b **Calculate the new budget for the handbags department. The first step has been done for you.**

Step 1: 1% of the handbag department's budget is £31000 ÷ 100 = £310

Step 2: 102.2% of the handbag department's budget is…

c **Calculate the new budget for the clothes department.**

You might want to start by finding 1% of last year's budget.

2 **Branches of a national chain of pound stores have been given income and expenditure budgets, outlined in Table 2.16.**

Table 2.16

	Stourport-on-Severn	Kidderminster	Dudley
Income budget	£1.8m	£2.1m	£1.7m
Actual income	£1.5m	£1.6m	£1.7m
Expenditure budget	£1.1m	£1.8m	£1.0m
Actual expenditure	£1.1m	£0.9m	£0.8m

a **How much profit were the three stores budgeted to make altogether?**

Step 1: subtract the expenditure budget from the income budget.

Stourport-on-Severn: £1.8m – £1.1m = £0.7m

Kidderminster: £2.1m – ____________ = ____________

Dudley: ____________ – ____________ =

Step 2: add them together to find the total profit budget.

b **How much profit did the three stores actually make?**

Step 1: Stourport-on-Severn: £1.5m – £1.1m = £0.4m

Kidderminster: £1.6m – ____________ = ____________

Dudley: ____________ – ____________ = ____________

Step 2: combine to find the total profit.

c **What was the profit variance for the three stores combined? State whether this was adverse or favourable.**

Step 1: calculate the difference between your answers to parts **a** and **b**.

Step 2: consider if this discrepancy means more or less profit for the business.

C Practice questions

3 A nightclub has an expenditure budget of £2500 and an income budget of £10 000 for a themed club night. The actual expenditure is £3200 and there is a £2560 favourable income variance.

a Calculate the actual income for the themed club night.
b Calculate the variance for the expenditure budget.
c Explain why the manager of the nightclub might not be too concerned about the answer to part **b**.

4 A manager is given an expenditure budget of £60 000. Her income budget is £96 000.

a Calculate her ratio of income budget to expenditure budget. Ensure the ratio is in its simplest form.
b She exceeds her income budget by 26%. Calculate the variance and indicate whether this is adverse or favourable.
c She exceeds her expenditure budget by 12%. Calculate the variance and indicate whether this is adverse or favourable.
d Calculate the variance in her profit budget.

5 A manager was given a £12 000 expenditure budget and a £19 000 income budget last year. This year he has an expenditure budget which is 3% higher and an income budget which is 11% higher.

a Calculate the manager's new expenditure budget.
b Calculate the manager's new income budget.
c The manger spends 40% of his new expenditure budget on staffing. How much, in pounds, does he spend on staffing?
d The manager secures a favourable variance in his profit budget of £3000, despite over-spending on his expenditure budget by £5000. Calculate the manager's actual income this year.

6 Table 2.17 outlines the budgets for various theatres, owned by the same company.

Table 2.17

Theatre location	Income budget	Actual income	Expenditure budget	Actual expenditure
London	£4m	£3.8m	£2.8m	£3m
Wells*	£8m	£5m	£0.9m	£2m
Manchester	£3.5m	£3.5m	£2m	£1.9m
Birmingham	£3.7m	£3.8m	£2.5m	£2.8m

*Wells is one of England's smallest cities.

a Calculate the variance in the profit budget for:
- **i** London
- **ii** Wells
- **iii** Manchester
- **iv** Birmingham

b Give two reasons why the manager of the Wells theatre should not be overly criticised for its profit variance.

Cash flow forecasting

Previous sections have looked at profit. Though profit is very important to a business, for many firms, cash flow is considered to be even more important. Having good cash flow means a business can pay those it owes money to on time. A business can be profitable but struggle with its cash flow and can therefore fail. Imagine a business selling very popular furniture. It allows its customers to buy now and pay in twelve months' time. In one year the business will have huge cash inflows. However, if it has not negotiated a long period of trade credit with its own suppliers, it could easily fail. The supplier of the furniture, its employees and the landlord of the store it operates from are likely to want to be paid now, not in twelve months. Though profitable, if the business has no cash it will not be able to continue.

Businesses and investors take cash flow very seriously. It is the life blood of a business. Because of this, firms often construct cash flow forecasts. These are predictions about the inflows and outflows of cash over time. Cash inflows refer to money coming in to the business, e.g. from sales to customers or from borrowed sources such as a bank loan. Cash outflows refer to money leaving the business, e.g. payments to suppliers or employees' wages. Looking ahead and anticipating any problems gives a business more time to take steps to avoid potential cash flow problems.

Please note there is an important difference between cash flow and profit. A bank loan of £1m improves a firm's cash flow tremendously. However, it does not directly improve its profit by £1m as the money must be repaid. Equally, delaying payments to suppliers improves cash flow, but does not improve profit as the money must still be paid to the firm's creditors.

Table 2.18 is an example of a cash flow forecast. Figures are in pounds.

Table 2.18

	January	February	March	April
Sales	1000	3000	5000	4000
Capital invested	7000	0	0	0
Total cash inflows	8000	3000	5000	
Raw materials	900	750	900	
Wages and other costs	600	800	1200	1100
Total cash outflows	1500	1550	2100	2000
Net cash flow	6500	1450	2900	
Opening balance	0	6500	7950	
Closing balance	6500	7950	10850	

This forecast looks at total cash inflows and total cash outflows. In this case, cash inflows come from sales of products and, in January, capital invested. This might be from the business owner's savings, for example.

Net cash flow is calculated by total cash inflows minus total cash outflows.

The firm's opening balance is the same as the previous month's closing balance.

The closing balance is calculated by adding net cash flow for the month, to the opening balance.

A Worked example

Use Table 2.18 to answer the following questions.

i Calculate total cash inflows for April.

4000 + 0 = £4000

ii Calculate the cost of the raw materials in April.
Total cash outflows were 2000. 1100 of this was wages and other costs, so the amount that must have been spent on raw materials was 2000 – 1100 = £900.

iii Calculate net cash flow for April.

4000 – 2000 = £2000

iv Calculate April's opening balance.

This is the same as March's closing balance, i.e. £10 850.

v Calculate April's closing balance.

Add the net cash flow for April of £2000 to its opening balance to give £12 850.

B Guided questions

Copy out the workings and complete the answers on a separate piece of paper.

1 Table 2.19 is a simplified cash flow forecast for a company. Note that the net cash flow for April is presented in brackets. In business, this is a standard way of indicating a number is negative, in this case −£15 000.

Table 2.19

	April	May
Total cash inflows	£20 000	£18 000
Total cash outflows	£35 000	£30 000
Net cash flow	(£15 000)	
Opening balance	£23 000	
Closing balance	£8 000	

a Calculate May's net cash flow.

Remember to take cash outflows away from cash inflows.

£18 000 – ____________ = ____________

b Calculate May's opening balance.

You will need to refer to the previous month to help you.

c Calculate May's closing balance.

Add May's net cash flow to its opening balance.

2 Table 2.20 gives another cash flow forecast.

Table 2.20

	July	August	September
Total cash inflows	£500	£400	£300
Total cash outflows	£500	£600	£900
Net cash flow	£0	(£200)	(£600)
Opening balance	£1700	£1700	£1500
Closing balance	£1700	£1500	£900

Calculate what would happen to September's closing balance if cash outflows that month were £500 higher than expected.

Step 1: calculate the new total cash outflows: £900 + £500 = ____________

Step 2: calculate the new net cash flow: £300 – ____________ = ____________

Step 3: the opening balance of £1500 would remain the same. Subtract your new net cash flow figure to get the revised closing balance.

C Practice questions

3 Calculate the missing values A to J in Table 2.21.

Table 2.21

	January £	February £	March £
Opening balance	0	**A**	**F**
Sales revenue	1000	**B**	4000
Bank loan	10000	0	0
Total cash inflows	11000	3000	**G**
Raw materials	2500	2000	2500
Wages	1500	1500	1600
Rent	900	900	900
Utilities	150	150	150
Debt interest and loan repayments	0	125	125
Total cash outflows	5050	**C**	**H**
Net cash flow	5950	**D**	**I**
Closing balance	5950	**E**	**J**

4 **Table 2.22**

	September £	October £	November £	December £
Opening balance	(20000)			
Total cash inflows	8000	10000	12000	13000
Raw materials	3000	3500	4000	4000
Wages	1500	1600	1700	1800
Rent and other costs	3000	3000	3000	3000
Total cash outflows	7500			
Net cash flow				
Closing balance				

a Using Table 2.22, calculate the closing balance for:

i September

ii October

iii November

iv December

b How would you describe the firm's cash flow?

c What, if any, action would you recommend the business take to improve its cash flow?

5 **Table 2.23**

	January £	February £	March £
Sales (Total inflows)	20000	30000	
Raw materials	20000	16000	17000
Wages and other costs	15000	12000	13000
Total cash outflows	35000	28000	30000
Net cash flow	(15000)	2000	
Opening balance	5000	(10000)	(8000)
Closing balance	(10000)	(8000)	(5000)

a Using the cash flow forecast in Table 2.23, calculate the sales for March.

b Re-calculate March's closing balance if February's raw materials were 20% more expensive than predicted.

6 Table 2.24 is a simplified cash flow forecast for April and May.

Table 2.24

	April	May
Total cash inflows	£2000	£3000
Total cash outflows	£1500	£2900
Net cash flow	£500	£100
Opening balance	£100	£600
Closing balance	£600	£700

Calculate how May's closing balance would change if May's total cash inflows were actually only 65% of the level expected and May's total cash outflows were 10% higher than initially forecasted.

Exchange rates

Exchange rates refer to one currency expressed in terms of another currency. These frequently fluctuate because of changes in demand and supply for the currencies in question. At the time of writing £1 = $1.56 and £1 = €1.42. When the exchange rate changes, there are consequences for businesses.

Imagine the pound appreciates (becomes stronger) against the dollar, i.e. it can now buy more dollars (and it costs more dollars to buy one pound). If a company imports its materials from America, a strong pound will be good news for them as their raw materials will cost less in pounds. However, if a business is selling its products to Americans (e.g. an American holidaymaker considering whether to holiday in the UK) the British products will be more expensive to customers buying in dollars.

A strong/appreciated pound makes British goods appear more expensive to foreign consumers and foreign goods appear cheaper to British business or customers.

A weak/depreciated pound makes British goods appear cheaper to foreign consumers and foreign goods appear more expensive to us.

A Worked examples

a The pound appreciates against the dollar. Formerly £1 = $1.56. Now £1 = $1.9. An American citizen wants to buy a British car. In pounds, it costs £14 500. In dollars, calculate the change in the cost for the American consumer of buying a British car because of the change in the exchange rate.

Step 1: previously £1 = $1.56

So £14 500 = $22 620 (Here you multiply $1.56 by the cost of the car in pounds. Both sides of the equation have been multiplied by £14 500.)

Step 2: now £1 = $1.9

So £14 500 = $27 550 (Again, both sides of the equation were multiplied by the cost of the car.)

The car now costs Americans $27 550 – $22 620 = $4930 more, despite the UK car company still receiving the same amount in pounds and not having raised its prices. This could easily be enough of an increase to encourage the American consumer to look into buying a car from a different country with a more favourable exchange rate instead.

b A firm in the UK imports components from an American company. A component costs the UK firm $45. In pounds, calculate the change in the cost of the component now the exchange rate has altered. Use the same exchange rate figures as in the previous worked example.

This is calculated in the same way as worked example **a**, but with an added stage. The question requires you to translate dollars into pounds so it would be useful to see what one dollar is equal to, in pounds. If both sides of the exchange rate equation are divided by 1.56, this is achieved:

Step 1: previously £1 = $1.56 or $1 = £0.64102…

So $45 = £28.846…

Step 2: now £1 = $1.9 or $1 = £0.52631… (both sides were divided by $1.9)

So $45 = £23.684…

The $45 component is now £5.16 cheaper to the British company in pounds (using unrounded figures) as the pound has appreciated. This is good for this firm as it will have lower costs. It is likely to benefit from higher profit margins while the pound remains stronger.

B Guided questions

Copy out the workings and complete the answers on a separate piece of paper.

1 Imagine £1 = ₹100 (Indian rupees). Calculate the cost, in rupees, of an Indian business importing the following things:

a Silver costing £2500

£1 = ₹100

£2500 = __________ (multiply the number of rupees by £2500)

b Financial services costing £14 000

£1 = ₹100

£14 000 = __________

c **Power generating equipment costing £2m**

- m is a standard abbreviation for million
- Remember: two million looks like this £2 000 000.

2 **Imagine £1 = ₹100 (Indian rupees). Calculate the cost, in pounds, of buying the following items when on holiday in India.**

a **Tour of Deli (₹400)**

£1 = ₹100. For this question it is more useful for you to know what ₹1 is worth. You can calculate this by dividing both sides of the equation by the number of rupees (₹100)

₹1 = £__________

₹400 = £__________

b **Evening meal (₹160)**

₹160 = £__________ (multiply the value of £1 (in rupees) by 160)

C Practice questions

3 A perfume business exports 2000 bottles of perfume to Japan. They cost £23 each. Assume £1 = ¥193.

a Calculate the cost in yen to the Japanese importer of British perfume.

b How is the Japanese importer affected if the exchange rate changes to £1 = ¥180?

c What might the consequences of your answer to part **b** be to the British exporter?

4 Bella has been on a business trip to Germany. She spent €2300 in total. When filing this for expenses, she does so in pounds. Calculate her expenses, in pounds, if £1 = €1.42.

5 A computing company imports most of its computers from a supplier in Thailand. Last year, each unit cost ฿6570 (Thai Bhat). This year, the Thai computer manufacturer raised its price by 12%. At the same time the Thai Bhat changed in value from £1 = ฿50 last year to £1 = ฿58 this year.

a Calculate the cost, in pounds, of a computer from the Thai supplier last year.

b Calculate the cost of a computer, in pounds, this year.

6 A business is choosing its new supplier of t-shirts. The firm's primary concern is price.

a Use Table 2.25 to work out which supplier offers the cheapest deal, in pounds.

Table 2.25

Supplier	Country	Price	Exchange rate
Supplier A	America	$4	£1 = $1.52
Supplier B	Turkey	₺8.5	£1 = ₺4.2
Supplier C	China	¥20	£1 = ¥9.70

b Give two factors, other than price, that the business should consider when selecting a supplier.

7 Table 2.26 is some data from an American business.

Table 2.26

Average cost per unit (including cost of goods sold and operating expenses)	$120
Operating profit margin	14%
Exchange rate	£1 = $1.60

Calculate the cost to a British consumer, in pounds, of importing one of this firm's products.

Liquidity ratios

This section will explain two liquidity ratios. Ensure that you are comfortable with the section on Ratios (page 10) before you look at this section.

Liquidity ratios allow a business to compare the current assets it has with its current liabilities. Current assets include cash in the bank and money owed to the business that is to be paid within 12 months. Inventory (or stock) is also a current asset. This refers to the financial value of raw materials or finished products that the business owns. Current assets also include cash or other assets expected to be converted into cash within a year.

Current liabilities are payments the business must make within one year. For example, if a business has a short-term loan that must be repaid in eight months' time, or a supplier to whom the business owes money in six months' time, these are both current liabilities. A business must ensure good liquidity to ensure survival.

There are two main liquidity ratios:

$$\text{current ratio} = \frac{\text{current assets}}{\text{current liabilities}}$$

$$\text{acid test ratio} = \frac{\text{current assets} - \text{inventory}}{\text{current liabilities}}$$

Worked example

A business has £78 000 worth of current assets (£68 000 of which is stock/inventory) and £50 000 of current liabilities.

i Calculate this firm's current ratio. Explain what this means for the business.

$$\text{current ratio} = \frac{£78\,000}{£50\,000} = 1.56$$

This means for every £1 of current liabilities, the business has £1.56 of current assets. This is a very promising figure.

ii Calculate this firm's acid test ratio. Explain what this means for the business.

$$\text{acid test ratio} = \frac{£78\,000 - £68\,000}{£50\,000} = 0.2$$

This means for every £1 of current liabilities, only £0.20 is held in liquid assets (i.e. current assets excluding stock).

Interpreting the figures

Though opinion varies, a current ratio of 1.5 is often considered to be ideal. This means the business should be able to comfortably cover its current liabilities. However, interpretation is not always this simple. As shown in the example on page 53, the business has a very good current ratio. However, the acid test ratio gives more information.

An ideal acid test ratio also varies but is commonly thought to be 1. The business in the previous worked example has a far lower figure. Why does this matter? A lot of this firm's current assets are tied up in stock, e.g. raw materials or finished goods. It may be difficult to sell these products. They might perish, or go out of date in the case of food, or go out of fashion in the case of clothes or technological products. This makes it difficult to transfer the value of the inventory owned into cash to repay debts.

Note that for this ratio, larger results are not always better. A very high figure for a liquidity ratio indicates the business has too many current assets and might be better off investing more in fixed assets to help generate more profit in the future. Though a business holding large amounts of cash may seem excellent, there is an opportunity cost of future investment.

A Worked example

A company has a current ratio of 1.7. It has current liabilities of £35 000. What is the firm's current assets figure?

This means the business has current assets which are 70% bigger than its current liabilities. So the business has current assets of £35 000 × 1.7 = £59 500.

If in doubt about the answer, put the numbers back into the original formula to check. In this case:

$$£59500 \div £35000 = 1.7$$

B Guided questions

Copy out the workings and complete the answers on a separate piece of paper.

1 **A business looks at part of its balance sheet (statement of financial position) shown in Table 2.27.**

Table 2.27

Cash	£12 000
Inventory	£10 000
Receivables	£5 000
Current liabilities	£25 000

a **Calculate this firm's total current assets.**

This is done by adding up the firm's assets which should be cash within twelve months. Calculate this by adding up the cash, the inventory (stock) and receivables (money owed to the business, to be paid within a year).

b **Calculate the firm's current ratio.**

$$\frac{\text{current assets}}{£25000} = \underline{\hspace{3cm}}$$

c **Explain what your answer to part b means.**

For every £1 of current liabilities the firm has…

d **Calculate the firm's acid test ratio.**

$$\frac{\text{current assets} - \text{inventory}}{£25000} = \underline{\hspace{3cm}}$$

e **Explain what your answer to part d means.**

For every £1 of current liabilities held…

2 **A business has various liabilities. It owes £30 000 to its suppliers. It has a short-term loan of £15 000 which must be repaid in less than 12 months and an overdraft of £5000. It also has a long-term loan of £60 000 that is due to be repaid in five years. Its current ratio is 0.5.**

a **Calculate this firm's current liabilities.**

Add up all the current liabilities mentioned. Remember not to include the long-term bank loan (this is a non-current liability as the business has over a year to repay).

b **Calculate this firm's current assets.**

Its current ratio of 0.5 tells you for every £1 of current liabilities the business only has 50p of current assets.

C Practice questions

3 A furniture company has a current ratio of 1.5 and an acid test ratio of 0.8.

a What explains the difference between the two liquidity ratios?

b Does this business have good liquidity?

4 Table 2.28 shows data for two competitors.

Table 2.28

	Company A	Company B
Current liabilities	£78 000	£54 500
Current assets	£81 400	£60 000

a Calculate the current ratio for company A.

b Calculate the current ratio for company B.

c Which company has the better liquidity, based on your findings in parts **a** and **b**?

d What further information might be useful to help you make judgements on the firms' liquidity?

5 Last year a business had current assets of £45 000 and current liabilities of £50 800. This year current assets have remained the same, however current liabilities have fallen by one fifth.

a Calculate the firm's current ratio for last year.

b Calculate the firm's current ratio for this year.

6 Using Table 2.29, assess which business has the best liquidity. Use ratios to support your answer.

Table 2.29

Business	Wiggly Ltd	Jiggly Ltd
Current assets including stock	£14 200	£19 650
Current liabilities	£8 000	£10 000
Stock	£4 000	£12 500

Return on capital employed and gearing ratios

Return on capital employed (ROCE) is a profitability ratio that allows people, e.g. managers or shareholders, to see how much operating profit was made as a result of the money invested in the business. Capital employed refers to the resources a business has available to it to help the firm generate its operating profit. It comprises money invested by shareholders, any long-term borrowing (non-current liabilities) and retained profits. These retained profits refer to past profits that shareholders allowed the business to keep for future business activities, rather than be paid out as dividends.

$$\text{return on capital employed} = \frac{\text{operating profit}}{\text{capital employed}} \times 100$$

ROCE is expressed as a percentage. The higher the return on capital employed percentage, the better. As a minimum, the ROCE should always be greater than the rate of interest the business could get for simply saving any capital in a bank. Managers might compare their ROCE figure with previous years, competitors' ROCE figures and their own objectives.

A Worked example

A car company has £17.5m capital employed and an operating profit of £2.92m. Calculate this company's return on capital employed.

$$\frac{£2.92}{£17.5\text{m}} \times 100 = 16.69\%$$

Gearing looks at the proportion of capital employed that is from borrowed sources. A business is usually described as 'highly geared' if it has a gearing ratio in excess of 50%. A business which has a low gearing ratio has a gearing ratio of less than 50%. A business with a high gearing ratio may find it harder to get future loans and is likely to be more vulnerable to rises in interest rates.

Gearing is calculated using the following formula:

$$\frac{\text{non-current liabilities}}{\text{capital employed}} \times 100$$

A Worked example

A car company has £17.5m capital employed. £12.7m of that is in the form of non-current liabilities. Calculate this company's gearing ratio.

$$\frac{£12.7\text{m}}{£17.5\text{m}} \times 100 = 72.57\%$$

B Guided questions

Copy out the workings and complete the answers on a separate piece of paper.

1 **Below is some financial data from a publishing firm:**

Operating profit	**£10.56m**
Total capital employed	**£114.15m**
Non-curent liabilities	**£5.5m**

a Calculate the publisher's return on capital employed.

$$\frac{\underline{\qquad\qquad}}{£114.15\text{m}} \times 100 = \underline{\qquad\qquad}$$

b Calculate the publisher's gearing ratio.

Remember that this ratio looks at the proportion of capital employed which comes from borrowed sources.

$$\frac{\underline{\qquad\qquad}}{£114.15\text{m}} \times 100 = \underline{\qquad\qquad}$$

2 **A business has a gearing ratio of 30% as it has £90 000 worth of non-current assets. It made £55 000 operating profit this year. Calculate the firm's return on capital employed.**

Step 1: calculate total capital employed. 30% of the capital employed is £90 000.

Step 2: refer back to the return on capital formula earlier in the section if you need to and calculate the ratio. Remember to express it as a percentage.

Practice questions

3 The data in Table 2.30 is for an insurance company this year and last year.

Table 2.30

	This year	Last year
Share capital	£10.4m	£9.0m
Retained earnings	£530.3m	£777.9m
Current liabilities	£29.7m	£24.5m
Non-current liabilities	£204.9m	£117m
Operating profit	£355.3m	£370.4m

a Calculate return on capital employed for last year.

b Calculate return on capital employed for this year.

c What was the firm's gearing ratio last year?

d What was the company's gearing ratio for this year?

e Has the firm's financial performance improved or worsened compared to last year?

f What might explain the change in financial performance?

4 A sole trader has £30 000 capital employed in her business and has a return on capital employed figure of 23%. Turnover was £45 000.

a Calculate the entrepreneur's operating profit.

b Calculate her operating profit margin.

5 A business has a gearing ratio of 25% and return on capital of 34%. Operating profit was £1.8m.

a Calculate capital employed.

b What is the size of this firm's non-current liabilities?

6 You are considering buying shares in one of these two companies.

Table 2.31

	Company A	Company B
Revenue	\$190 700m	\$250 800m
Gross profit	\$92 000m	\$123 950m
Overheads	\$63 800m	\$75 630m
Capital employed	\$113 400m	\$130 060m

a Using appropriate profitability ratios, explain which company you would prefer to invest in. (Refer to the sections on Ratios and Profit margins.)

b Give two examples of other pieces of data that might be useful in helping you decide which business to buy shares in.

Efficiency ratios

Efficiency ratios allow businesses to measure how efficiently they manage certain assets and liabilities. This section will look at three key efficiency ratios: receivables days, payables days and inventory turnover.

Receivables refers to money that is owed to the business, usually from customers who have bought goods on credit (buy now, pay later). For a business selling to other businesses, receivables days of around 28 days is very common. This means businesses wait approximately 28 days to be paid by their customers. For firms selling to consumers, allowing customers to buy goods on credit is common in some industries involving large purchases, e.g. buying a new car or furniture, but is rare in others, e.g. having a meal in a restaurant. As the name suggests, this ratio is expressed as a number of days.

The formula for calculating receivables days (also known as debtor days) is:

$$\text{receivables days} = \frac{\text{receivables}}{\text{revenue}} \times 365$$

Ⓐ Worked example

A large stationery company has receivables of £1 900 800 and sales turnover of £16 540 300. Calculate the firm's receivables days.

$$\frac{£1900800}{£16540300} \times 365 = 41.95 \text{ days}$$

This means that, on average, this stationery company must wait just under 42 days to receive the money that it is owed by customers.

To give this number more meaning, it is often compared to the payables days figure. Payables (or creditor) days refers to the time it takes the business to pay the money it owes to its suppliers. A large payables days figure can be beneficial as it can assist the firm with its cash flow. Of course, delayed payment should always be done with permission from the supplier. Ideally, a business will have a payables days figure higher than its receivables days figure. This means the business can receive cash from its customers before having to pay its suppliers.

Payables days (creditor days) is calculated using the following formula:

$$\text{payables days} = \frac{\text{payables}}{\text{cost of sales}} \times 365$$

A Worked example

The stationery company discussed in the previous worked example has a payables figure of £1 790 400 and a cost of goods sold of £14 591 200. Remember 'cost of goods sold' is often used interchangeably with the term 'cost of sales'. Calculate its payables days.

$$\frac{£1790400}{£14591200} \times 365 = 44.79 \text{ days}$$

When compared to the firm's receivables days, the stationers is in a pleasing situation. On average, customers pay the business more quickly than the business pays its suppliers.

The final efficiency ratio is inventory turnover. Also known as stock turnover, this ratio helps a business calculate how quickly its inventory is sold. Average inventories held can be calculated by the opening stock (value of the inventory at the start of the period) plus the closing stock, divided by two. Inventory is usually measured at cost (how much it cost the business to buy rather than how much the business can sell it on for).

The ratio is expressed as a number of times. For example, a ratio of 3 means that the average inventory is sold 3 times over the year. This ratio is calculated using the following formula:

$$\text{inventory (stock) turnover} = \frac{\text{cost of goods sold}}{\text{average inventory held}}$$

Here, the higher the figure the better. It indicates the firm is able to sell its stock quickly which is good for its cash flow. An acceptable figure varies from industry to industry. For example, a firm selling perishable products such as food would need to have a higher inventory turnover figure than, say, a DIY store which has a large range of products which do not tend to go off quickly or become outdated.

Inventory turnover can also be expressed as a number of days by using the following formula:

$$\frac{365}{\text{inventory turnover}} = \text{number of days}$$

A Worked example

A large stationery company has a cost of goods sold of £14 591 200 and average inventories held of £2 918 240.

i Calculate its inventory turnover. Explain what this figure means.

$$\frac{£14591200}{£2918240} = 5 \text{ times}$$

This means that, on average, the business sells its inventory five times in one year.

ii Now express its inventory turnover as a number of days.

$$\frac{365}{5} = 73 \text{days}$$

This means the business sells its inventory every 73 days.

B Guided questions

Copy out the workings and complete the answers on a separate piece of paper.

1 Table 2.32 gives some financial data from a bakery chain in Scotland.

Table 2.32

Revenue	£803 960
Receivables	£26 890
Average inventory held	£15 790
Cost of sales	£320 700
Payables	£89 900

Using this information, calculate:

a the difference between the firm's receivables days and payables days

Step 1: calculate the bakery's receivables days:

$$\frac{£26890}{£803960} \times 365 = \underline{\hspace{3cm}}$$

Step 2: calculate its payables days:

$$\frac{}{£320700} \times 365 = \underline{\hspace{3cm}}$$

Step 3: find the difference between the two and comment on the value of the difference.

b the firm's stock turnover

$$\frac{\text{cost of goods sold}}{£15790} = \underline{\hspace{3cm}}$$

2 At the start of the year a business had inventory worth £12 000. By the end of the year it had inventory worth £16 000. Cost of goods sold was £126 000. In days, calculate the firm's inventory turnover.

Step 1: calculate average inventory held by adding the inventory at the start of the year with the inventory held at the end of the year and dividing by two.

Step 2: use the inventory turnover formula:

$$\frac{£126\,000}{\text{average inventory held}} = \underline{\hspace{3cm}}$$

Step 3: translate this into the number of days by using this formula:

$$\frac{365}{\text{inventory turnover}} = \underline{\hspace{2cm}} \text{ days}$$

C Practice questions

3 A large homeware business has a USP of offering customers a wide range of choice in terms of colour and design. It has an inventory of £356m at the start of the year and £340m at the end of the year. Its cost of goods sold is £368m.

a Calculate the firm's inventory turnover figure.

b Express its inventory turnover as a number of days.

c What might explain its inventory turnover figure?

d Why might this inventory turnover figure pose a problem for the business?

4 Table 2.33 gives information about a private limited company.

Table 2.33

Revenue	£55 000
Receivables	£38 000
Total current assets	£90 000
Cost of sales	£20 000
Payables	£8 500
Total current liabilities	£70 000

a Calculate the firm's current ratio. Refer back to the section on Liquidity ratios (page 53) if you need to.

b Calculate the firm's payables days.

c Calculate the firm's receivables days.

d How well is the firm managing its liquidity? What steps might you recommend to help improve the situation?

5 A CEO calculates her business has an inventory turnover of 10 times. Cost of sales is £60 000. What is average inventory held?

6 Table 2.34 shows some financial information for a sole trader from last year. It also shows how this information has changed this year.

Table 2.34

	Last year	Percentage change compared with last year
Revenue	£145 000	–3.5%
Cost of sales	£30 000	–5%
Average inventory held	£12 000	+30%
Receivables	£105 000	+15%
Payables	£22 500	+20%

a Calculate the difference between the firm's payables days, last year and this year.
b Calculate the difference between the firm's receivables days, last year and this year.
c Is your answer to part **b** a concern?
d What is the difference, in days, between the firm's inventory turnover last year and this year?

Investment appraisal

Businesses need to ensure future investments are worthwhile. They must deliver a good return for shareholders. Firms often have more than one option of how to spend any retained profit or borrowed funds and need to ensure they spend it in the most profitable way. For example, a supermarket may consider developing its online retailing, open smaller convenience stores or enter a new territory such as America. It is unlikely to be able to afford to do all of these at the same time: a choice must be made.

This section looks at three investment appraisal tools that businesses can use to help them make decisions about which projects to pursue. All three methods involve looking at forecasted returns and costs of the project, and therefore answers calculated are estimates and are only as useful as the reliability of the data entered into the calculation. A business entering a very different market might find it difficult to accurately predict the full future costs and revenues of such a move.

Payback

The payback period refers to the length of time a project will take to make the amount of money spent on it. Ideally, an investment will have recouped any costs as soon as possible. A business with particular concerns over cash flow, or impatient short-termist shareholders, is likely to pay close attention to this investment appraisal tool.

A Worked example

A business is considering investing in new machinery. There will be an initial cost of purchasing the machinery of £100 000. Then, every year, the machine will need maintenance costing £5000. However, the machine is expected to generate £22 000 revenue annually. Calculate the payback period.

To calculate payback, it is easiest to present this information in a table. See Table 2.35.

The first column refers to the end of the year. Year 0 is usually used to describe the year in which the investment takes place, e.g. the machinery is bought and installed.

Cash inflows are zero in this year but are £22 000 every year thereafter.

Cash outflows refers to the cost of the machinery in year 0 and then the subsequent annual cost of the maintenance.

Net cash flows are calculated by the cash inflows – cash outflows of each year.

Remember numbers in brackets represent negative numbers. This information is needed for all investment appraisal methods.

Payback also requires an additional column: cumulative cash flow. This involves adding up the net cash flows over the years. It starts at negative £100 000. This negative number gets smaller and smaller because the net cash flow is positive. Eventually, by the end of year 6, the investment has paid for itself, the cumulative cash flow has become positive.

Table 2.35

End of year	Cash inflows	Cash outflows	Net cash flow	Cumulative cash flow
0	£0	£100 000	(£100 000)	(£100 000)
1	£22 000	£5 000	£17 000	(£83 000)
2	£22 000	£5 000	£17 000	(£66 000)
3	£22 000	£5 000	£17 000	(£49 000)
4	£22 000	£5 000	£17 000	(£32 000)
5	£22 000	£5 000	£17 000	(£15 000)
6	£22 000	£5 000	£17 000	£2 000

For this project it can be seen that payback occurs after five years and a certain number of months — somewhere in the sixth year payback occurs and the cumulative cash flow figure reaches zero.

Businesses often want to know exactly where in that sixth year payback occurs. This is calculated as follows:

$$\frac{\text{amount still to pay back}}{\text{net cash flow for the year in which payback occurs}} \times 12$$

In this case, by the end of year 5, £15 000 is required to reach payback. Within the sixth year the business will receive £17 000.

$$\frac{£15000}{£17000} \times 12 = 10.59 \text{ months}$$

So payback occurs after 5 years and 10.59 months.

A common student mistake is to simply write that payback occurs after 10.59 months — don't forget to include the number of years as well.

Please note that this method assumes that the cash inflows and outflows are evenly spread throughout the year, which in reality is unlikely.

Average rate of return

Payback looks at how quickly a project recoups its cost. For many businesses this is useful, but not all they want to know. Firms also want to know the profitability of an investment, i.e. how much of a return they get compared to the money spent on it. In some cases, it might be worth taking a one year longer payback period to secure a higher profit. Average rate of return (ARR) is expressed as a percentage. Businesses want the ARR to be as high as possible and certainly above the interest rate the business could get from simply saving money in a bank account.

Calculating ARR is not mathematically difficult but does have a few steps to it so is worth practising.

Step 1: add up the project's total profit (remember to take away any costs).

Step 2: divide the project's profit by the number of years you have considered, e.g. the life span of the machine. Do not include year 0 here.

Step 3: divide your answer in step 2 by the initial size of the investment.

Step 4: multiply your answer to Step 3 by 100 to get your ARR percentage. This number means (on average) the percentage of the initial investment that the business will get back every year for the life time of the project.

In summary, the ARR formula is:

$$\frac{\text{average yearly profit}}{\text{cost of investment}} \times 100$$

A Worked example

Refer back to the table from the previous worked example. The important column for ARR will be the net cash flow. Calculate the ARR for this investment.

Over the six years of the investment in the machinery the business makes:

$$(100000) + £17000 + £17000 + £17000 + £17000 + £17000 + £17000 = £2000 \text{ profit}$$

As an average this is $£2000 \div 6 = £333.33\ \ldots$

Compared to the initial investment, the ARR for this business is:

$$\frac{£333.33\ldots}{£100000} \times 100 = 0.33\%$$

This ARR is very small and the project would almost certainly be rejected.

Net present value

Net present value (NPV) is the only investment appraisal method studied here that takes into account the time value of money. This refers to the idea that money received now is more valuable than money received years in the future. If given the choice between getting £1000 today or £1000 next year, it is logical to receive the money now. This money could be invested and generate a return by next year. If you wait a year to receive the money, then there is a chance that it will not be paid to you. Additionally, goods and services usually become more expensive over time so £1000 will not be able to buy you as many things in a year's time. For this reason, you might even be willing to accept, say £950 today, instead of £1000 a year from now.

Because net present value takes into account *when* the money is received by the business, and not just how much is received, a discounted net cash flow needs to be calculated using discount factors (also known as discount rates). Discount factors can be in various sizes. A business will use a larger discount rate if they want to place a lower value on money received in the future compared with now. A discount rate of 5% is shown in Table 2.36.

Table 2.36

Year	Discount rate (5%)
0	1
1	0.952
2	0.907
3	0.864
4	0.823
5	0.784
6	0.746

To calculate net present value, multiply each year's net cash flow by the discount rate for that year to get the present value for the net cash flow. Then add up all the present values to get the net present value figure, expressed in pounds (or whichever currency the investment is in).

A Worked examples

a **Find the total net present values from Table 2.37.**

Table 2.37

End of year	Net cash flow	Discount factor (5%)	Present value
0	(£100 000)	1	(£100 000)
1	£17 000	0.952	£16 184
2	£17 000	0.907	£15 419
3	£17 000	0.864	£14 688
4	£17 000	0.823	£13 991
5	£17 000	0.784	£13 328
6	£17 000	0.746	£12 682

Adding up the net present values gives a total of (£13 708).

Businesses would usually reject a project that is expected to give a negative net present value. A positive number may be accepted and the larger the figure the better.

b **Tables 2.38 and 2.39 are cash flow tables for two possible projects:**

Table 2.38 Project A

Year	Net cash flow (£)	Cumulative cash flow (£)	Discount factor (5%)	Discounted net cash flow (£)
0	(50 000)	(50 000)	1	(50 000)
1	0	(50 000)	0.952	0
2	0	(50 000)	0.907	0
3	0	(50 000)	0.864	0
4	100 000	50 000	0.823	82 300

Table 2.39 Project B

Year	Net cash flow (£)	Cumulative cash flow (£)	Discount factor (5%)	Discounted net cash flow (£)
0	(60 000)	(60 000)	1	(60 000)
1	25 000	(35 000)	0.952	23 800
2	25 000	(10 000)	0.907	22 675
3	25 000	15 000	0.864	21 600
4	25 000	40 000	0.823	20 575

i Calculate payback, ARR and NPV for both projects.

Project A's payback

Payback occurs after three years and...

$$\frac{£50\,000}{£100\,000} \times 12 = 6 \text{ months}$$

Project B's payback

Payback occurs after two years and...

$$\frac{£10\,000}{£25\,000} \times 12 = 4.8 \text{ months}$$

Project B has the shorter payback period.

Project A's ARR

Profit = £100 000 – £50 000 = £50 000

Average profit per year = £50 000 ÷ 4 = £12 500

$$\text{ARR} = \frac{£12\,500}{£50\,000} \times 100 = 25\%$$

Project B's ARR

Profit = £100 000 – £60 000 = £40 000

Average profit per year = £40 000 ÷ 4 = £10 000

$$\text{ARR} = \frac{£10\,000}{£60\,000} \times 100 = 16.67\%$$

Project A has the higher (therefore better) ARR.

Project A's NPV

Here the total discounted net cash flow gives a total net present value of £32 300 after subtracting the initial £50 000 investment.

Project B's NPV

Adding up the discounted net cash flows (including subtracting the initial investment of £60 000) gives a total net present value of £28 650.

ii Which project would you recommend the business pursue? What other factors should be considered?

- Though project B has the better payback period, project A has a higher NPV and ARR so project A is probably preferable on financial grounds.
- However, the business would be likely to consider more than the results from the investment appraisal. Project A could be problematic for a firm's cash flow as there are no cash inflows until year 4. However, it does have the advantage of having smaller start-up costs compared to project B.
- A business may also consider qualitative factors: what do its employees think about the projects? Do the current workforce have the required skills? How might the investment affect the firm's reputation?

B Guided questions

Copy out the workings and complete the answers on a separate piece of paper.

1 **A business is considering an investment. The cash inflows and outflows are shown in Table 2.40. Calculate this project's payback period.**

Table 2.40

End of year	Cash inflows	Cash outflows	Net cash flow	Cumulative cash flow
0	£0	£200 000	(£200 000)	(£200 000)
1	£50 000	£10 000	£40 000	(£160 000)
2	£80 000	£10 000		
3	£120 000	£10 000		

Step 1: complete the net cash flow column. Remember, this is done by subtracting cash outflows from cash inflows.

Step 2: complete the cumulative cash flow column by adding the net cash flow of year 2, to the cumulative cash flow of year 1 and so on.

Step 3: work out the year in which cumulative cash flow must have turned positive.

Step 4: calculate the exact number of months using this formula:

$$\frac{\text{amount still to pay back}}{\text{net cash flow for the year in which payback occurs}} \times 12$$

2 **A company is considering whether to open a number of new stores at an initial cost of £700 000. Net cash inflows are expected to be £200 000 annually for the first two years and £250 000 a year for years 3 and 4.**

Calculate the average rate of return for opening new stores.

Step 1: calculate the total profit for the venture. Remember to take the initial cost of opening the stores away from the net cash inflows for years 1 to 4.

Step 2: divide this total profit by the number of years being analysed (four).

Step 3: divide this average profit figure by the initial cost of the investment.

Step 4: multiply your answer by 100 to get your ARR percentage.

3 **A private limited company is considering whether to open a new factory. The cash inflows and outflows are detailed in Table 2.41 along with a 10% discount factor. Using this, calculate the project's net present value and consider whether the new factory should be opened.**

Table 2.41

End of year	Cash inflows	Cash outflows	Net cash flow	Discount rate	Present value
0	£0	£150 000	(£150 000)	1	
1	£45 000	£2 000	£43 000	0.909	
2	£47 000	£2 000		0.826	
3	£50 000	£2 000		0.751	
4	£54 000	£2 000		0.683	

Step 1: complete the net cash flow column by subtracting cash outflows from cash inflows.

Step 2: multiply the discount rate by the net cash flow figure, for each year.

Step 3: add up the total present values (including subtracting the initial investment cost from year 0).

Step 4: decide whether the project is viable, according to the NPV tool. Remember that you are looking for a positive number.

C Practice questions

4 The possible investment options in Table 2.42 are presented to a board of directors. Which would you recommend they go ahead with and why?

Table 2.42

	Investment Option 1	Investment Option 2	Investment Option 3
Initial cost of investment	£300 000	£100 000	£900 000
Payback period	3 years, 1 month	2 years	4 years, 6 months
Average rate of return	9%	5%	7%
Net present value (5% discount rate used)	£20 000	£15 000	£50 000

5 Table 2.43 gives details on returns from two potential projects.

Table 2.43

End of year	Project A's net cash flow	Project B's net cash flow
0	(£500 000)	(£100 000)
1	£200 000	£30 000
2	£200 000	£47 000
3	£200 000	£55 000
4	£200 000	£60 000

a Using the information in the table, calculate the payback period for the two projects.
b Calculate the ARR for both projects.
c Which project would you recommend the business invests in?
d What other factors might the business take into account, other than the payback period and ARR?

6 Upon the strong recommendation of its new finance director, a business invested in expensive state-of-the-art technology for all factories. The director had presented the figures in Table 2.44 to the board who were impressed with its potential for an excellent return. Unfortunately, the project did not deliver what it had promised.
a Calculate the difference between the predicted ARR and the actual ARR.
b Explain what might account for such a difference in figures.

Table 2.44

End of year	Predicted net cash flow	Actual net cash flow
0	(£1.1m)	(£1.1m)
1	£0.5m	£0.1m
2	£0.7m	£0.2m
3	£0.9m	£0.4m
4	£1.2m	£0.5m

7 The human resources manager believes the company should invest heavily into staff training at a cost of £850 000. She believes this would lead to annual net cash flows of £250 000 for 4 years.

Table 2.45

End of year	Discount factor (5%)	Discount factor (10%)
0	1	1
1	0.952	0.909
2	0.907	0.826
3	0.864	0.751
4	0.823	0.683

a Calculate the net present value of this investment using:
  i the 5% discount factor in Table 2.45
  ii the 10% discount factor in Table 2.45

b Explain why the NPV figures gained in your answers to parts **ai** and **aii** are different.

c To what extent do you think the staff training should go ahead?

3 Marketing

Market share, size and growth

This section largely comprises percentages and percentage change calculations. You may wish to revisit the pages covering these topics (pages 12 and 16) before studying this section.

Managers often want to know what is happening to the size of the market/industry in which they operate. This knowledge can help entrepreneurs decide whether or not to enter a certain market segment. It can also help managers see if their sales are growing in line with the growth of the industry as a whole.

The size of a market can be measured in two different ways: by volume or by value. Measuring the market size by volume looks at the number of units sold, e.g. the number of litres of fruit juice sold in the UK in one year.

Usually, market size is measured by value: the total revenue earned from selling a product, e.g. the fruit juice market being worth over a billion pounds.

A Worked example

Last year, 2500 million litres of bottled water were sold in the UK at an average price of £0.70 per litre.

i What is the market size by volume?

This is 2500m litres of bottled water.

ii What is the market size by value?

This is calculated by the number of litres sold, multiplied by the price per litre sold, i.e. $2500\text{m} \times £0.7 = £1750\text{m}$

Decision makers not only want to look at the size of a market, but also how much it is changing by. A market can be very large but if it is shrinking quickly, many new businesses would choose not to enter it. The rate at which a market is increasing in size is known as market growth and is calculated by doing a percentage change. The formula is:

$$\frac{\text{new market size} - \text{original market size}}{\text{original market size}} \times 100$$

A Worked example

Last year, 2500 million litres of bottled water were sold in the UK at an average price of £0.70 per litre. This year, 2550 million litres of bottled water were sold in the UK. The average price per litre rose to £0.71. Calculate the market growth in terms of value, in the UK bottled water market this year, compared with last year.

As calculated in the previous worked example, the value of the bottled water market last year was £1750m.

This year the new market size is:

$$2550\text{m} \times £0.71 = £1810.5\text{m}$$

The market growth is therefore:

$$\frac{£1810.5\text{m} - £1750\text{m}}{£1750\text{m}} \times 100 = 3.46\%$$

As well as wanting to know market size and growth, businesses often want to know the proportion of sales in the market that *they* gained. The higher a firm's market share, the better. This is calculated as a percentage. The market share formula is:

$$\frac{\text{sales of one company}}{\text{market size}} \times 100$$

Worked example

Last year, the UK bottled water market was worth £1750m. This year it is worth £1810.5m. The market leader had £350m worth of sales last year and £398.31m sales this year. Calculate the market leader's market share last year and this year.

Last year's market share:

$$\frac{£350\text{m}}{£1750\text{m}} \times 100 = 20\%$$

This year's market share:

$$\frac{£398.31\text{m}}{£1810.5\text{m}} \times 100 = 22\%$$

Guided questions

Copy out the workings and complete the answers on a separate piece of paper.

1 **The energy drinks market increased in size from 450m litres last year, to 490m litres this year. The market leader saw its market share, in terms of volume, fall from 25% to 23% over the same time period.**

a **Calculate the market growth in the industry.**

$$\frac{£490\text{m} - \_\_\_\_}{\_\_\_\_} \times 100 = \_\_\_\_\_\_\_\_$$

b **Calculate the change in the number of litres sold by the market leader, this year, compared with last year.**

Step 1: calculate the market leader's market share last year.

Last year the market leader sold 25% of 450m litres.

1% = 4.5m litres (divide market size by 100)

25% = 112.5m litres (multiply by 25)

Step 2: calculate the market leader's market share this year.

This year the market leader sold 23% of 490m litres.

1% = ____________

23% = ____________

Step 3: calculate the change in the number of litres the market leader sold.

2 Imagine the chart shows the market share for pizza takeaways for the UK, last year.

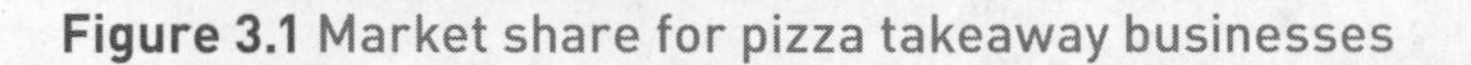

Figure 3.1 Market share for pizza takeaway businesses

a Pizza Pie had £150m of sales last year. Calculate the size of the market, by volume.

7.5% = 150m

1% = ______________ (divide £150m by 7.5 to get 1% of the total market size)

100% = ______________

b Calculate the size, in pounds, of Wow Pizza's sales.

Multiply the value of 1% of the market by Wow Pizza's market share.

c On average, each pizza sold for £9.50. Calculate the market size by volume.

Divide the market size in value by the average price of a pizza, to get the number of pizzas sold.

C Practice questions

3 Using Table 3.1, find out which is the largest market, by value, this year.

Table 3.1

Market	Market size last year, by value	Market growth this year
A	£800m	+2%
B	£700m	+12%
C	£990m	–19%
D	£200m	+120%

4 A market sells 300 000 units at an average price of £1100.

a Calculate the market size in value.

b The next year the market sells 10% more units despite raising prices by 5%. Calculate the new market size by value.

c As a percentage, calculate how much the market has grown, in value, over the last year.

5 Table 3.2 gives market shares for firms operating within the confectionery industry.

Table 3.2

Business	Sales value (£m)	Market share
Sweet and Spice	1 680	
Cocoa	1 050	10%
Sugarlicious	630	
Sweetie Pie		5%
Others		
Total	10 500	100%

a Calculate Sweet and Spice's market share.

b Calculate Sugarlicious' market share.

c Calculate Sweetie Pie's sales value.

d Calculate the sales value and market share for the other firms in the industry.

6 A business had 5% market share by volume last year when it sold 64 000 units. This year it sold 90 000 units and has a 6% market share.

a Calculate, as a percentage, the market growth by volume in this market over the last year.

b Has the firm's own sales growth been quicker or slower than the market growth? Use calculations to support your answer.

7 A company has 5% of the footwear market, 1% of the childrenswear market and 12% of the womenswear market. It has an operating profit margin of 30%. Calculate this firm's operating profit using Table 3.3.

Note: operating profit and profit margins are explained in Unit 2 Finance.

Table 3.3

Market	Size (£m)
Womenswear	15 500
Childrenswear	4 500
Footwear	5 200

Price elasticity of demand

When a business increases its prices, demand for its products (all other things staying the same) will fall — and vice versa for a price decrease. Businesses find it very useful to know how *much* demand will fall when prices rise. This helps them decide how to set their prices in order to maximise revenue.

Price elasticity of demand is a measure of how sensitive customers are to changes in price. It is calculated using the following formula:

$$\frac{\text{percentage change in quantity demanded}}{\text{percentage change in price}}$$

Revisit Unit 1 Key mathematical skills if you need further guidance on how to do a percentage change (page 16).

Usually, the price elasticity of demand (PED) figure is negative. This is because if prices rise (+) quantity demanded will fall (–). If prices fall (–) quantity demanded will rise (+). This inverse relationship results in a negative PED sign for most products. The PED figure tells businesses whether their products have price elastic or price inelastic demand.

Price elastic demand indicates customers are sensitive to changes in price. A small rise in price will lead to a more than proportional drop in sales and therefore fall in revenue. PED figures of less than −1 indicate price elastic demand, e.g. −1.5 or −4 (note that no units are required here).

Price inelastic demand refers to customers being unresponsive to changes in price. A price rise of 10%, for example, will lead to a fall in sales of less than 10%. The PED figure will be less than zero but greater than −1 e.g. −0.5 or −0.1. For businesses with price inelastic demand, revenue is usually maximised when prices are increased.

A Worked examples

a When a local cinema increases its ticket prices by 10% it notices sales of tickets fall by 5%. Calculate this firm's PED and explain what this means.

$$\text{PED} = \frac{-5}{10} = -0.5$$

This means demand for this cinema's tickets is price inelastic. For every 1% rise in price, quantity demanded falls by only 0.5%.

b When a local ice skating rink raised its prices from £11 to £12.32, the number of customers coming to skate fell by 15%. Calculate the PED of the ice skating rink and explain what this means.

Here, the percentage change in price must be calculated first.

$$\frac{£12.32 - £11}{£11} \times 100 = 12\%$$

Now the PED formula can be used:

$$\frac{-15}{12} = -1.25$$

This indicates demand for the ice skating rink is price elastic. Customers *are* responsive to price changes. For every 1% rise in price, quantity demanded falls by 1.25%.

The formula can also be rearranged to calculate the likely change in the quantity sold following a price change.

c A business has a price elasticity of demand figure of −2.5. The business lowers prices by 14%. Calculate what will happen to the number of units sold, as a result of the price change.

The percentage change in quantity demanded = PED figure × price change

i.e. $-2.5 \times -14 = 35\%$ increase in sales

B Guided questions

Copy out the workings and complete the answers on a separate piece of paper.

1 **A business decreases prices from £4 to £3.50. Consequently, demand rises from 1200 units sold to 1500 units sold. Calculate the price elasticity of demand for this firm's products.**

Step 1: calculate the percentage change in the firm's quantity demanded.

$$\frac{1500-1200}{\quad} \times 100 = \_\_\_\_\_\_$$

Step 2: calculate the percentage change in its price.

$$\frac{£3.5-}{£4} \times 100 = \_\_\_\_\_\_$$

Step 3: refer back to the PED formula. Divide the percentage change in quantity demanded by the percentage change in price.

2 **A manager decides to raise her prices from £20 to £25. Demand for the products has a price elasticity figure of −2. Calculate what is expected to happen to quantity demanded.**

Step 1: calculate the percentage change in the firm's price.

Step 2: multiply your answer to Step 1 by the PED figure.

Practice questions

3 A manager is comparing the price elasticity of demand of goods made by two firms. The results are shown in Table 3.4.

Table 3.4

	Company A's product	Company B's product
Percentage change in price	−16%	−2%
Percentage change in quantity demanded	+5%	+5%

a Calculate the price elasticity of demand for company A.

b Calculate the price elasticity of demand for company B.

4 A business doubles its prices. As a result, sales drop from 80 000 to 75 000 units. Calculate this firm's price elasticity of demand figure and explain its meaning.

5 In a meeting, three junior managers are trying to estimate the price elasticity of demand of a product. Matilda believes the product will have price inelastic demand of −0.5 because of the company's excellent branding. Amy estimates a PED of −1.5 due to the competitive nature of the market and there being many alternatives available for customers. Uwais believes a PED of −1 is more likely.

Upon an increase in price from £8 to £10, sales fall by one third.

a Calculate the firm's actual PED.

b Who was closest with their elasticity estimate?

c Sales volumes were initially 60 000. Calculate the new number of units sold.

d How has revenue changed due to the price rise?

6 A business sells 30 000 units for £2.50 each one year. The next year it increases its prices to £6. The price elasticity of demand for its products is −0.2.
  a Calculate the percentage increase in the firm's price.
  b Calculate the new number of units sold after the price rise.
  c Calculate how the firm's revenue changes due to the price rise.

Income elasticity of demand

Businesses often like to be able to anticipate how their sales will be affected if their customers' income changes. Economic occurrences such as recessions can mean that, on average, customer income falls. Some businesses do well out of such events, others do very badly. Equally, customer income may rise, e.g. due to an economic boom, changes in national minimum wage laws or changes in income tax. Income elasticity of demand (YED) helps businesses make predictions about how they will be affected if customer income changes.

The formula for income elasticity of demand is:

$$\frac{\text{percentage change in quantity demanded}}{\text{percentage change in customer income}}$$

Businesses must interpret the sign of the YED figure (+ or −) and the value of the number itself. Table 3.5 shows how figures are interpreted.

Table 3.5

Figure	Interpretation
A positive sign	The product is a normal or luxury good. Customer income has a direct relationship with quantity demanded. As customers get richer, they buy more of these sorts of products. Most products would have a positive sign.
A negative sign	This product is an inferior good. People buy more of it the poorer they are, e.g. bus travel and value brands. They switch away from these products as their income grows.
Absolute value is greater than one, e.g. +2 or −1.5	Regardless of the sign (+ or −) this product has income elastic demand. A small change in income leads to a large change in demand for the product.
Absolute value is less than one, e.g. +0.3 or −0.7	Whether positive or negative, this product has income inelastic demand. A change in income does not have a substantial impact on sales.

Some examples are shown in Table 3.6.

Table 3.6

YED figure	Meaning
−2	An inferior good with income elastic demand.
+0.3	A normal good with income inelastic demand. Sometimes referred to as normal necessities, these products might include toothpaste and toilet roll. You might buy more of these products and better brands if you are richer, but not a great deal of any extra income gets spent on these sorts of products.
+1.9	This product has income elastic demand and is a normal good. These products are often referred to as luxury goods. Examples include top of the range TVs, foreign holidays and sports cars.

A Worked examples

a In a local area, average incomes fall by 15%. As a result, a local pay-weekly company sees a rise in sales of 20%. (Pay-weekly businesses often sell household products such as televisions, furniture and washing machines. They allow customers to pay in weekly installments for these goods, instead of having to buy the product outright before taking it away.) Calculate the YED for this company's products and explain your answer.

$$\text{YED} = \frac{20}{-15} = -1.33$$

This company is selling inferior goods with income elastic demand.

This business will do better the more its customers' incomes fall.

This fits with expectations as poorer customers are more likely to opt to pay weekly for goods rather than pay in full up front.

As customers become richer, they would switch away from this method of payment as, in the long run, paying weekly costs more because interest is charged.

b A business sells products with a YED of +1.5. Following a rise in customer incomes of 5%, what would you expect to happen to quantity demanded?

The change in quantity demanded = percentage change in income × YED value

Quantity demanded would rise by 5 × 1.5 = 7.5%

Guided questions

Copy out the workings and complete the answers on a separate piece of paper.

1 The owner of a small chain of curry takeaway shops wants to estimate the income elasticity of demand for her products. She notices that when average incomes in the area fall by 5%, demand for her takeaways rises from 40 curries per night to 46 per night. Calculate the income elasticity of demand for her takeaways.

Step 1: calculate the percentage change in demand for the curries. Revisit Percentage change on page 16 if you need additional help with this.

$$\frac{46-40}{} \times 100 = \underline{\qquad\qquad}\%$$

Step 2: use the income elasticity of demand formula:

$$\frac{}{-5} = \underline{\qquad\qquad}$$

2 A budget hotel company charges £35 for a night's stay. The marketing manager estimates the hotel rooms have a YED of −0.4. Last year the company had 54 000 nights booked at its hotels. This year, recession has hit and customer incomes have fallen, on average by 6%. The business keeps prices the same. Calculate the firm's change in revenue because of the recession.

Step 1: calculate the firm's original revenue (price × quantity sold).

Step 2: work out the new quantity demanded.

The change in quantity demanded = percentage change in income × YED value.

In this case: $-6 \times -0.4 = 2.4\%$ increase in quantity demanded because of the fall in income.

So, increase last year's quantity demanded (54 000 nights) by 2.4%.

Step 3: multiply the new number of nights sold by the nightly price to get the new revenue.

__________ × £35 = __________

Step 4: calculate the difference in the original and the new revenue figures.

C Practice questions

3 A rise in average incomes locally from £25 800 to £26 100 leads to a business selling 3% more units.

a Calculate this firm's income elasticity of demand.

b Explain what your answer to part **a** tells you about the firm's products.

4 A business sells a range of products. Its YED figures are outlined in Table 3.7.

Table 3.7

Product 1	−2
Product 2	−0.1
Product 3	+0.6
Product 4	+3

a Which product would you expect to do best in a recession?

b Which product would you expect to see the greatest rise in sales in prosperous economic times, e.g. an economic boom?

c Which product is likely to be least affected by any changes in customer income?

d Customer incomes rise by 20% and one of the products sees a fall in quantity demanded of 2%. Which product would this be?

e Why might businesses want to sell a range of products like these?

5 A bus company estimates its bus travel service has an income elasticity of demand figure of −0.56. Bus passenger numbers rise by 5% compared to the previous year.

a All other things remaining the same, what has happened to customer income?

b What other factors could explain the rise in bus passenger numbers?

6 The sales figures for two sports car companies are outlined in Table 3.8.

Table 3.8

	Company A	Company B
Number of cars sold last year	4 000	650
Number of cars sold this year	4 200	689

Over the past year, customer income rose by 3%.

a Calculate the income elasticity for company A's cars.

b Calculate the income elasticity for company B's cars.

c Which company's cars have the most income elastic demand?

Forecasting

It is useful for firms to be able to make reliable predictions about the future. Often businesses particularly want to be able to forecast their sales growth.

This knowledge can help inform decisions about whether to expand, downsize, enter other markets and so on. Often businesses use a range of methods to help make forecasts. This section will look at moving averages, extrapolation and correlation. These methods all involve looking at past data. Very new businesses will often find such techniques hard as they have little historical data to look back on.

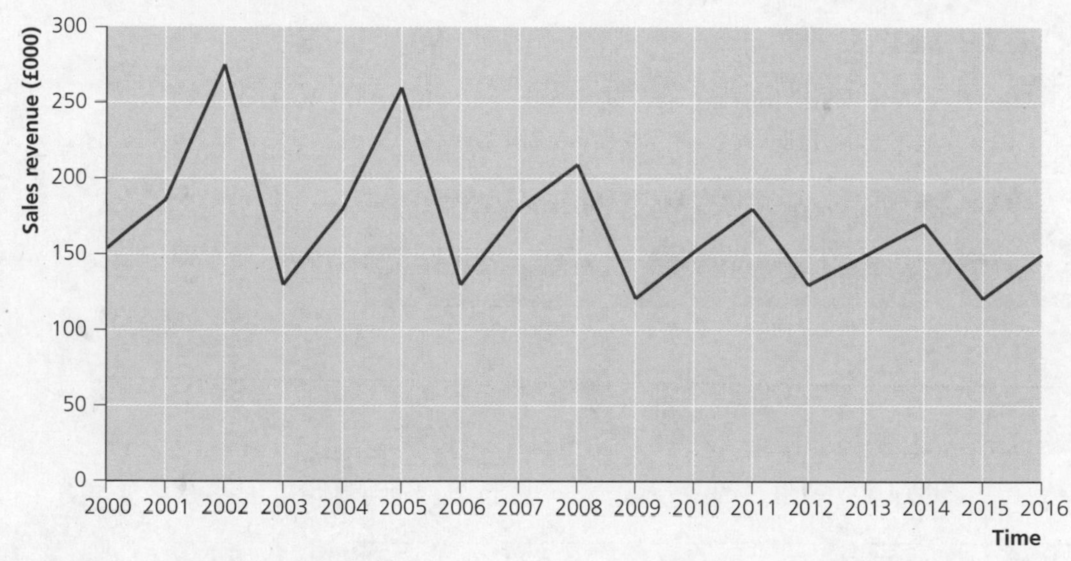

Figure 3.2 Sales revenue

Figure 3.2 illustrates the fluctuations in a firm's sales revenue. From this, it is difficult to make long-term predictions about the future sales revenue for the company. It is clear that sales turnover fluctuates, but the rate, overall, at which it is increasing or decreasing is unclear. Moving averages can be useful here to smooth out data so the underlying trend can be seen. Revisit Averages on page 6 before studying the next section, if required.

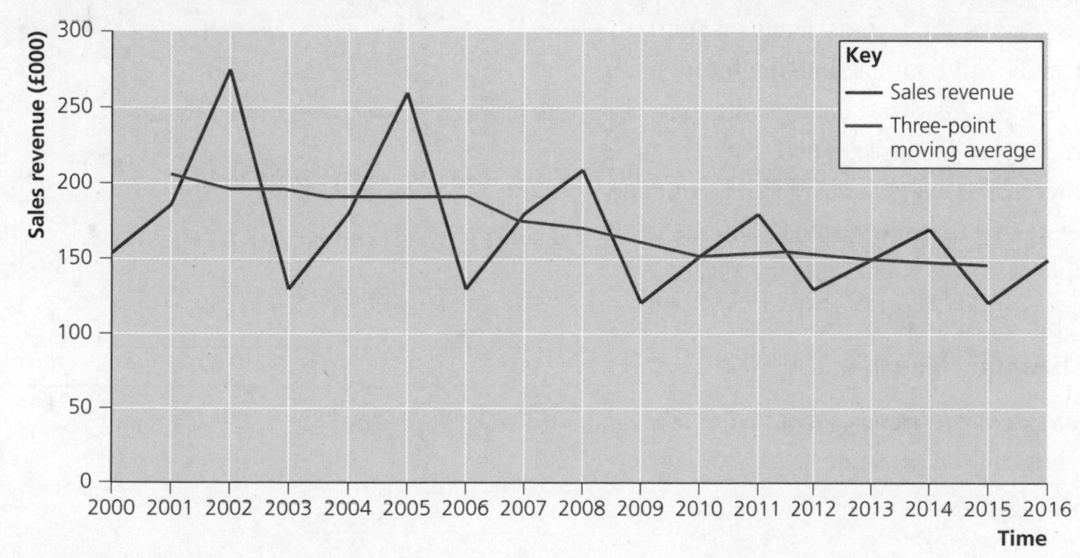

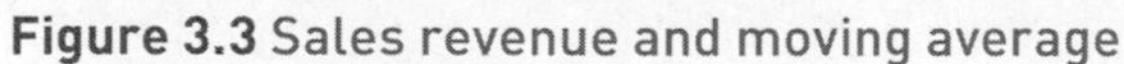

Figure 3.3 Sales revenue and moving average

Figure 3.3 shows the original sales revenue and a three-point moving average. It is now easier to see that there is a downward trend in this firm's sales revenue, once fluctuations have been smoothed out. This information can tell firms they need to take action in order to boost sales revenue.

Three-point moving averages are calculated by adding the first three pieces of data, e.g. sales for 2000, 2001 and 2002, and dividing by three. Then an average is created for data from 2001, 2002 and 2003. 2000's sales revenue is no longer part of this average and 2003's data has joined the average. See the worked examples below.

A Worked example

Table 3.9 shows how the first few years' moving averages were calculated for Figure 3.3.

Table 3.9 Calculating moving averages

Year	Sales revenue (£000s)	Three-year total (£000s)	Three-year moving average (£000s)
2000	154		
2001	186	154 + 186 + 275 = 615	$\frac{615}{3} = 205$
2002	275	186 + 275 + 130 = 591	$\frac{591}{3} = 197$
2003	130	275 + 130 + 180 = 585	$\frac{585}{3} = 195$
2004	180	130 + 180 + 260 = 570	$\frac{570}{3} = 190$
2005	260		

Another sales forecasting technique is extrapolation. This assumes that current trends will continue into the future and so, using historical data, predictions are made. For example, if sales have been increasing every year by 3%, it could be expected that sales would continue to increase at this rate, into the future.

This idea can be expressed on a chart as shown in Figure 3.4.

Extrapolated trends should be treated with caution. It is unusual for trends to continue in the same way forever.

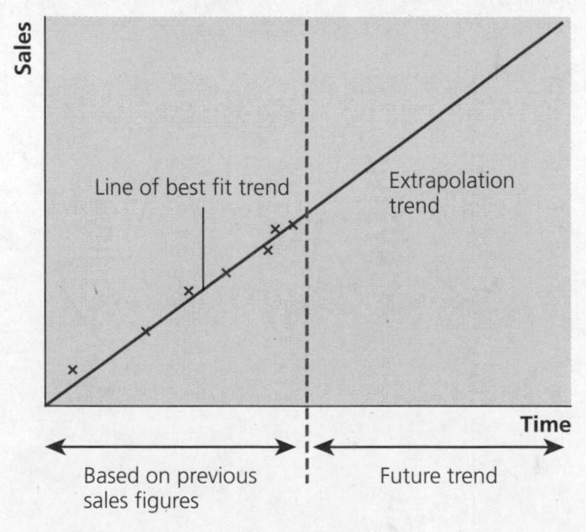

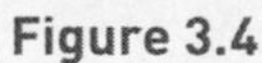

Figure 3.4

A Worked example

Table 3.10 shows the revenue of a large chain of coffee shops. Each year, revenue has grown by 8%. By extrapolating the trend, calculate the revenue you would expect the chain to have in 2017 and 2018.

Revisit Unit 1 Key mathematical skills if you need further help with percentage changes.

Table 3.10

Year	Sales revenue (£m)
2014	200
2015	216
2016	233.28
2017	
2018	

2017 forecasted sales revenue:

1% of 2016's sales revenue is 2.3328

108% = £251.9424m (£251.94m to 2 decimal places)

2018 forecasted sales revenue:

For 2018, look back at 2017's sales revenue. Use the unrounded answer.

1% of 2017's sales revenue is 2.519424

108% = £272.1m

Correlation

Correlation allows businesses to see the relationship between two variables.

Positive correlation indicates that values increase together or decrease together. For example, a business might notice that the more it spends on advertising, the more products it sells. The more it spends on staff training, the higher quality might be.

Negative correlation occurs when one value increases and the other decreases. This is known as an inverse relationship. Most businesses would see a negative correlation between the price they charge and the amount they sell. If they increase prices, sales will fall and vice versa.

Correlation also looks at the **strength** of a relationship. Strong correlation indicates a close link between a change in one value and a corresponding change in another. A product with price elastic demand would see a strong negative correlation between its price and quantity sold — a rise in price would lead to a substantial fall in the quantity demanded.

Correlation can also be weak — a change in one factor leads to a relatively small change in another. Again, referring to price elasticity, a business selling a product with price inelastic demand would notice weak negative correlation between its selling price and quantity sold. Imagine a cigarette manufacturer raises its prices. Some smokers would cut back on their consumption but, due to the addictive nature of the product, there would be only a small drop in sales.

A business with highly paid workers might notice a weak positive correlation between employee pay and output per worker (i.e. labour productivity). If the employees are already well paid, extra money may make a little difference to the amount of effort they put into their work, but only cause a slight change.

A Worked examples

a Correlation can be shown on a graph. The correlation between spending on advertising and sales revenue for a business is shown in Figure 3.5. A line of best fit has also been drawn here. This can be done by eye or by using computer software, such as Excel, to draw it more accurately.

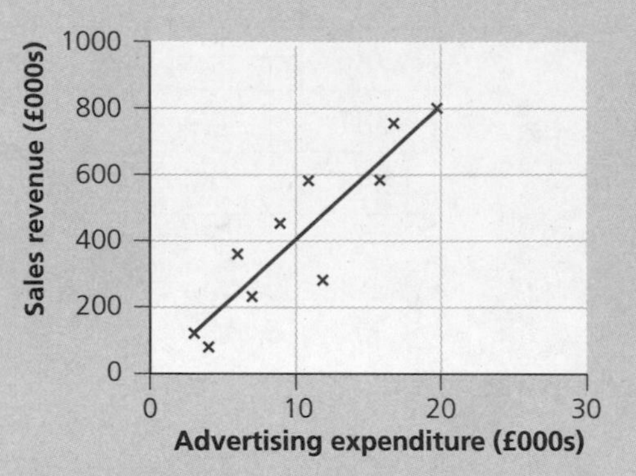

Figure 3.5

i Describe the relationship between advertising expenditure and sales revenue.

- There is a strong, positive correlation between spending on advertising and sales revenue.
- Using the line of best fit, it can be seen that for every £10 000 increase in advertising expenditure, sales revenue increases by £400 000.

ii Considering this relationship, what would you recommend the business does regarding its advertising expenditure?

- The business should invest more heavily in advertising so it can maximise its sales revenue as advertising appears to be highly influential in this industry.
- Spending £10 000 on advertising leads to £400 000 extra sales revenue which is likely to be a profitable move.

b The idea of correlation can be applied to various business concepts, not just marketing. Here is an operations example.

A factory is trying to raise the productivity of its employees (increase output per worker). The human resources manager suggested giving employees more breaks. The operations manager thought the opposite should be done – give fewer breaks to allow more time for working on the factory floor.

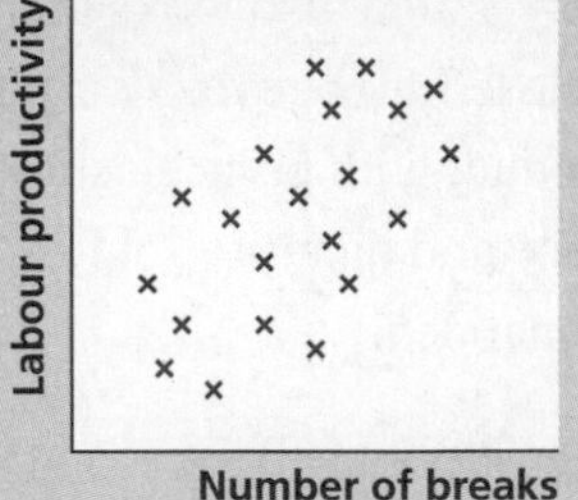

Figure 3.6

i Describe the correlation between breaks and productivity.

There is a weak positive correlation here — the higher the number of breaks, the higher labour productivity is.

ii Who was correct? The human resources manager or the operations manager?

The Human Resources Manager — giving more breaks is correlated with higher productivity.

iii Would you expect the correlation on the chart to exist regardless of the number of breaks?

No. After a certain number of breaks, productivity would probably fall as there is not enough time to work.

B Guided questions

Copy out the workings and complete the answers on a separate piece of paper.

1 Table 3.11 shows the number of cars sold per year, by a small car dealership.

Table 3.11

Year	Number of cars sold	3-year total	3-point moving average
2010	160		
2011	210	520	173.33
2012	150	490	163.33
2013	130	480	
2014	200		
2015	180		
2016	160		

a Calculate the three-point moving average for 2012 to 2014.

Remember to divide the three-year total by 3.

b Calculate the three-point average for 2013 to 2015.

Step 1: add up the sales of cars for 2013, 2014 and 2015.

Step 2: divide your total by 3.

c Now do the same to calculate the three-point moving average for 2014 to 2016.

2 A business had a consistent percentage increase in its sales volumes for the past three years. Use Table 3.12 to extrapolate the trend to forecast sales volumes in 2017 and 2018.

Table 3.12

Year	Sales volumes (000s)
2014	120
2015	126
2016	132.3

Step 1: identify the growth in the sales volumes each year, currently. Remember the percentage change formula:

$$\frac{\text{change in the values}}{\text{original value}} \times 100$$

Step 2: for 2017, increase 2016's figure by the percentage found in Step 1. You may find it easiest to find 1% of 2016's sales volume first by dividing by 100:

1 323 000 ÷ 100 = 1323

Step 3: repeat Step 2 for 2018's sales volume forecast, adding the trend rate of growth to 2017's forecast.

C Practice questions

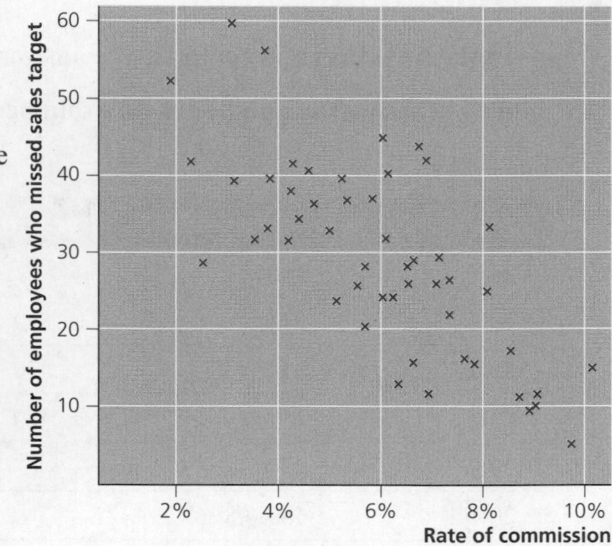

Figure 3.7 Effectiveness of commission

3 The sales director is analysing data from across her stores to see how effective commission for sales people is at generating sales. Each store has 100 salespeople. Commission involves paying salespeople a percentage of the selling price of the goods they sell. She plots data regarding the number of employees who miss sales targets per store and the rate of commission paid, on the scatter graph in Figure 3.7.

- **a** What correlation, if any, is shown on the scatter graph?
- **b** What explains the relationship seen?
- **c** Despite the scatter graph, why might the sales director be reluctant to introduce 10% commission in all stores?

4 Copy Table 3.13 and complete boxes A to D.

Table 3.13

Year	Sales revenue (£)	3-year total	3-point moving average
2011	500 000		
2012	550 000		A
2013	780 000		B
2014	520 000		C
2015	480 000		D
2016	650 000		

5 Figure 3.8 represents a market's size over recent years.

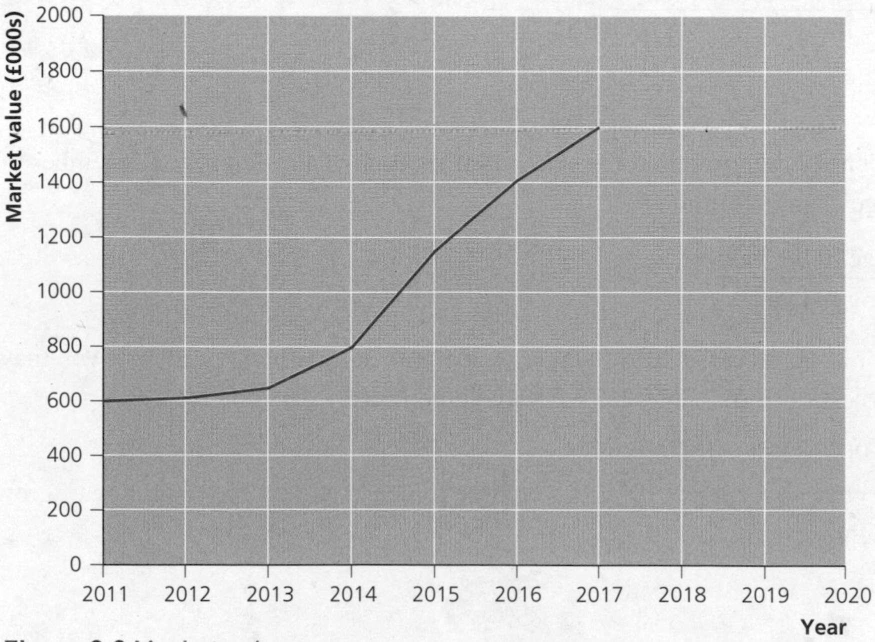

Figure 3.8 Market value

a By visually extrapolating the trend, what would you estimate the market size to be by 2020?

b Why might your answer to part **a** not be correct?

6 A manager and her team at a local travel agents have been trialing the use of Twitter as a way of getting more customers into the agency and selling more luxury holidays. She plots on a scatter graph the number of tweets done per week and the number of holidays sold that week, as shown in Figure 3.9.

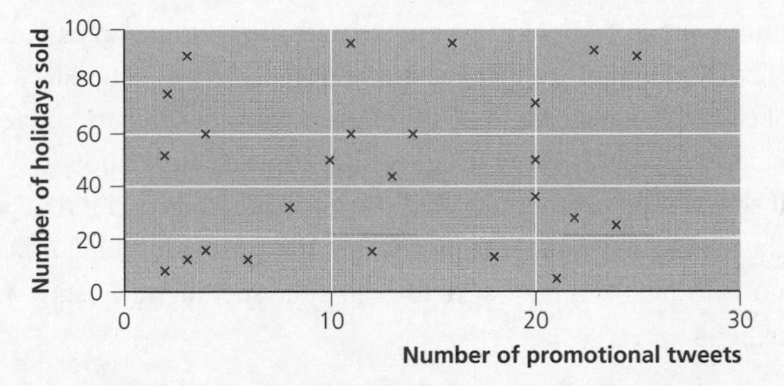

Figure 3.9

a How would you describe the correlation between the number of tweets and the number of holidays sold?

b Would you recommend the travel agents do more or fewer tweets in the future?

c What might have positive correlation with sales of luxury holidays?

7 **Table 3.14**

	Market size by value (£m)	
Year	**Market A**	**Market B**
2014	160	200
2015	176	202
2016	193.6	204.02

If market A and market B shown in Table 3.14 continue to grow at the same rate as they have done for 2014 to 2016, by 2018:

a What will market A's market size be?

b What will market B's market size be?

c Which will be the larger market?

4 Operations

Capacity and capacity utilisation

Capacity refers to the maximum output a business can make with current resources. This could be that a factory can make 25 000 cupcakes in a week, that a bus can seat 72 passengers or that a hotel can have a maximum of 15 guests. Capacity utilisation refers to the proportion of a firm's capacity that is being used. A high capacity utilisation is beneficial to a business as it shows high demand for the business's products and it reduces unit costs. This is because the fixed costs remain the same, but there are additional units of output, causing average costs to fall. You may wish to revisit the section on Costs (page 26) before reading further.

Expressed as a percentage, capacity utilisation is calculated using the following formula:

$$\frac{\text{total output}}{\text{maximum capacity}} \times 100$$

Firms typically prefer a high capacity utilisation figure. However, most businesses do not aim for 100% capacity utilisation because it reduces their flexibility, e.g. it makes it difficult to correct errors and take on new orders. It also does not allow for maintenance of machinery, staff training etc.

A Worked examples

a A hotel has 27 rooms. Due to a sporting event in the local area, capacity utilisation is much higher than normal and 26 out of the 27 rooms have been booked. What is this hotel's capacity utilisation?

$$\frac{26}{27} \times 100 = 96.3\%$$

b A business has variable costs of £2.50 per unit and fixed costs of £30 000 per month. The firm has a maximum capacity of 25 000 units per month. Last month, 10 000 units were made. This month 22 000 units were made.

i Calculate the firm's capacity utilisation and cost per unit last month.

Step 1: capacity utilisation $= \frac{10000}{25000} \times 100 = 40\%$

Step 2: cost per unit $= \frac{(10000 \times 2.5) + 30000}{10000} = £5.50$

ii Calculate the firm's capacity utilisation and cost per unit this month.

Step 1: capacity utilisation $= \frac{22000}{25000} \times 100 = 88\%$

Step 2: cost per unit: $\frac{(22000 \times 2.5) + 30000}{22000} = £3.86$

B Guided questions

Copy out the workings and complete the answers on a separate piece of paper.

1 A business earned £70 000 revenue last year from selling the products it manufactured for £20 each. The factory has capacity of 4000. Calculate the firm's capacity utilisation.

Step 1: calculate how many units the business made. In this case, this is calculated by dividing the firm's revenue by the price charged:

$$\frac{70000}{£20} = \_\_\_\_ \text{ units}$$

Step 2: now calculate the firm's capacity utilisation:

$$\frac{\_\_\_\_}{4000} \times 100 = \_\_\_\_ \%$$

2 Here is some information on a business's output and capacity utilisation. Calculate the percentage change in this firm's capacity this year, compared with last year.

Table 4.1

Year	Number of units made	Capacity utilisation
Last year	8925	85%
This year	9430	82%

Step 1: calculate last year's capacity.

85% of capacity = 8925 units

1% of capacity = 105 units

100% of capacity = 10 500 units

Step 2: calculate this year's capacity.

82% of capacity = 9430 units

1% = __________

100% = __________

Step 3: calculate the percentage increase or decrease in your answers to Step 1 and Step 2. Remember the percentage change formula is:

$$\text{percentage change} = \frac{\text{change in the values}}{\text{original value}} \times 100$$

Practice questions

3 Table 4.2 gives information on two bottling plants in the UK. Calculate the capacity utilisation for:

a bottling plant A

b bottling plant B

Table 4.2

Type of drink	Bottling plant A Number of drinks bottled	Bottling plant B Number of drinks bottled
Cola	0	1m
Water	1m	0.5m
Lemonade	0.5m	0.75m
Other	0.3m	0
Maximum capacity	2m	3.5m

4 Table 4.3 shows selected financial and operations data for a business.

Table 4.3

Fixed costs per year	£100 000
Variable costs per unit	£5.60
Maximum capacity per year	900 000

a Calculate cost per unit at:

i 30% capacity utilisation

ii 60% capacity utilisation

iii 100% capacity utilisation

b What explains the trend in your answers for parts **ai** to **aiii**?

5 **Table 4.4**

	Factory 1	Factory 2	Factory 3
Number of units made	5 830	4 263	3 180
Capacity utilisation	55%	87%	30%

With reference to Table 4.4, calculate the capacity of:

a Factory 1 **b** Factory 2 **c** Factory 3

Decision trees

A decision tree is one tool managers may use when making decisions, e.g. over which market to enter, whether to invest in a new product line or advertise existing products more heavily etc. Like investment appraisal, decision trees look at the costs and revenues from a project. However, unlike the investment appraisal methods studied, decision trees consider the probability of projects being a success and different potential outcomes from the same decision. Look at Figure 4.1.

- A square represents a decision the business must make, in this case whether to launch product A or product B, or do nothing and launch no further products.
- Written below the product A and product B arrows is the cost of the project. You will see these are presented in brackets.
- Circles represent a possible outcome from taking a decision. In this case it is possible there is high or low demand. The number below these arrows, for example, 0.7, represent the chance of that outcome occurring. These probabilities can be presented in one of two ways. As a proportion of 1 (as in this case): 0.7 indicates a 70% chance of high demand being the outcome of launching product A. 0.3 (or 30%) is the chance of low demand. When probabilities are presented in this way, they should equal one, when added together.

- Alternatively, probabilities can be expressed as a percentage: if this is the case, the probabilities should add up to 100%. For example, for product B, there is a 60% chance of high demand and therefore revenues of £12m but a 40% chance of low demand and gaining revenues of only £4m.
- From the decision tree, expected values (also known as expected monetary values, EMV) can be calculated. Expected values for a decision are calculated by multiplying the probability of each outcome by the revenue. All the expected values for a decision are then added together for each option.
- Finally, the cost associated with the decision is subtracted to give the net gain (also known as the net expected value.) The higher the figure, the better for the business and the more likely it will select that option.

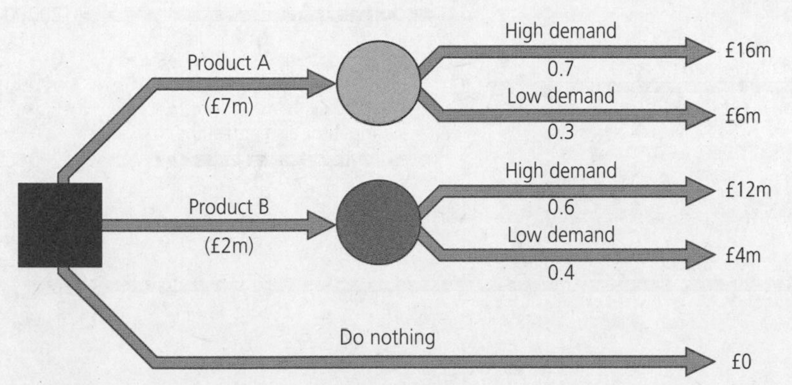

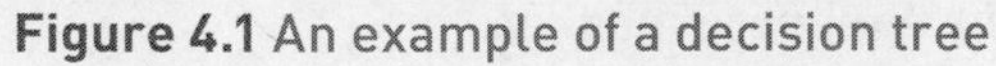
Figure 4.1 An example of a decision tree

A Worked example

Using the decision tree in Figure 4.1, which product has the highest net gain?

Product A

Step 1: high demand: £16m × 0.7 = £11.2m

Step 2: low demand: £6m × 0.3 = £1.8m

Step 3: total expected value: £11.2m + £1.8m = £13m

Step 4: now subtract the cost of product A to get project A's net gain: £13m − £7m = £6m

Product B

Step 1: high demand: £12m × 0.6 = £7.2m

Step 2: low demand: £4m × 0.4 = £1.6m

Step 3: total expected value: £7.2m + £1.6m = £8.8m

Step 4: now subtract the cost of product B to get project B's net gain: £8.8m − £2m = £6.8m

- Product B has the higher expected monetary value and, according to the decision tree, should be the chosen option.
- Additionally, it has a lower initial cost.
- However, it could be considered the slightly riskier project as project A has a greater chance of high demand than product B.
- Both options are more profitable than launching no additional products, assuming the figures in the decision tree are correct.

B Guided question

Copy out the workings and complete the answers on a separate piece of paper.

1 **Figure 4.2 is a decision tree relating to entering two different markets.**

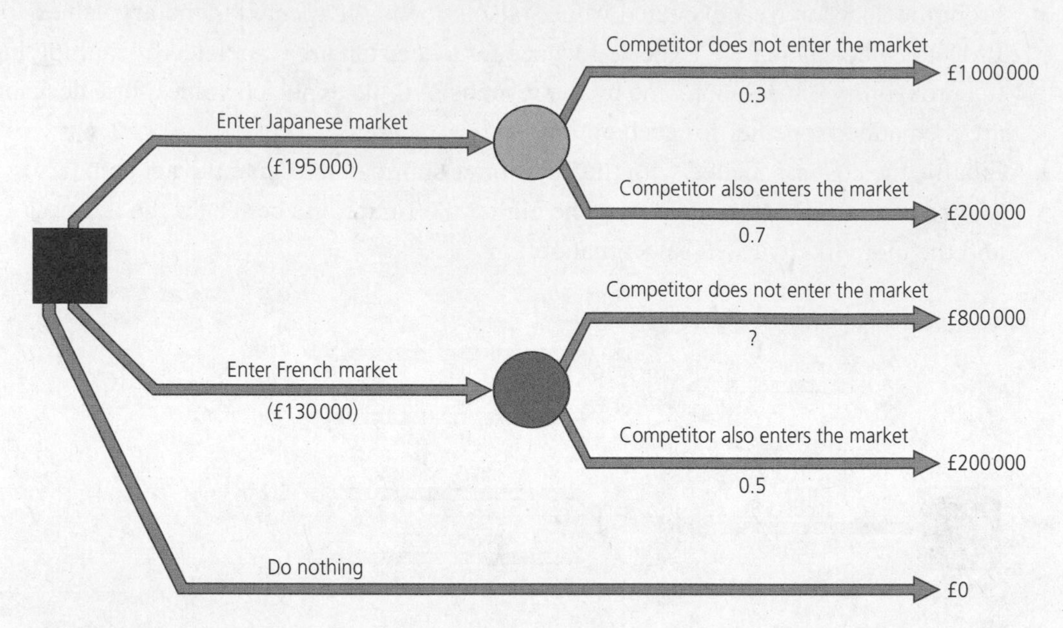

Figure 4.2

a Calculate the net gain of entering the Japanese market.

Step 1: multiply the probability of the competitor not entering the market (0.3) by the revenue gained from this outcome.

Step 2: multiply the probability of the competitor entering the Japanese market (0.7) by the revenue gained from this outcome.

Step 3: add your answers to Step 1 and Step 2 together to get the total expected value.

Step 4: subtract the cost of entering the Japanese market (£195 000) from your answer to Step 3 to find the net gain of this decision.

b Calculate the probability of a competitor not entering the French market. Express your answer as a percentage.

- The probabilities of the competitor entering the market and the competitor not entering the market must add up to one.
- From the decision tree it can be seen that there is a 0.5 probability of the competitor entering the market.
- To convert the decimal probability into a percentage probability, multiply your decimal probability by 100.

c Calculate the net gain of entering the French market.

- Follow the same steps as for part **a**.
- Remember net expected value and net gain have the same meaning.
- Don't forget to subtract the cost of entering the French market.

Practice questions

2 A London-based business is considering whether to advertise on the radio in the London area only or conduct a national TV advertising campaign, as it can sell products through its website to the whole of the UK. Figure 4.3 is a decision tree outlining the marketing manager's predicted costs, revenues and risks.

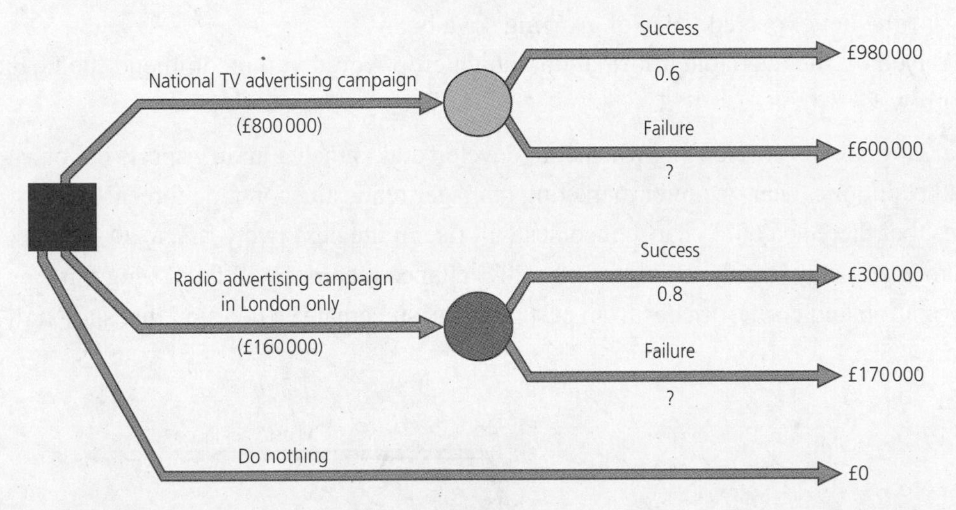

Figure 4.3

- **a** Calculate the probability of the national TV advertising campaign failing. Express your answer as a percentage.
- **b** Calculate the probability of the radio advertising campaign in London failing. Express your answer as a percentage.
- **c** Calculate the total expected value of the national TV advertising campaign.
- **d** Calculate the net gain from the national TV advertising campaign.
- **e** Calculate the total expected value of the London radio advertising campaign.
- **f** Calculate the net gain for the London radio advertising campaign.
- **g** Which decision would you recommend the business takes? Justify your view.

3 A farmer is deciding which crop to grow next year.

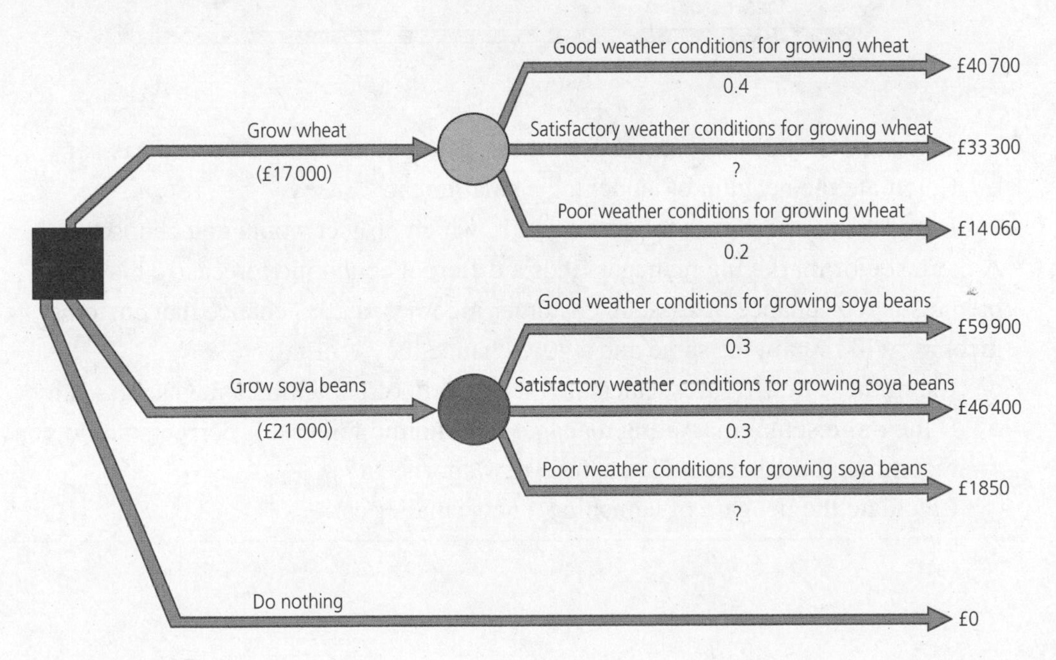

Figure 4.4

a Using the decision tree in Figure 4.4, calculate:
   i the probability of satisfactory weather conditions for growing wheat
   ii the probability of poor weather conditions for growing soya beans
   iii the size of the profit/loss the farm would make if soya beans were grown in poor weather conditions
   iv the net expected value of growing wheat
   v the net expected value of growing soya beans

b Based on the available information, which crop would you recommend the farmer grows?

4 A car company is considering whether to develop and launch a luxury sports car or a small economical car. A junior marketing manager reads an economic forecast which states that there is a 60% chance incomes will rise in the next two years, a 20% chance incomes will stay largely the same and a 20% chance incomes will fall. Using this information and cost estimates from past projects, she creates a decision tree shown in Figure 4.5.

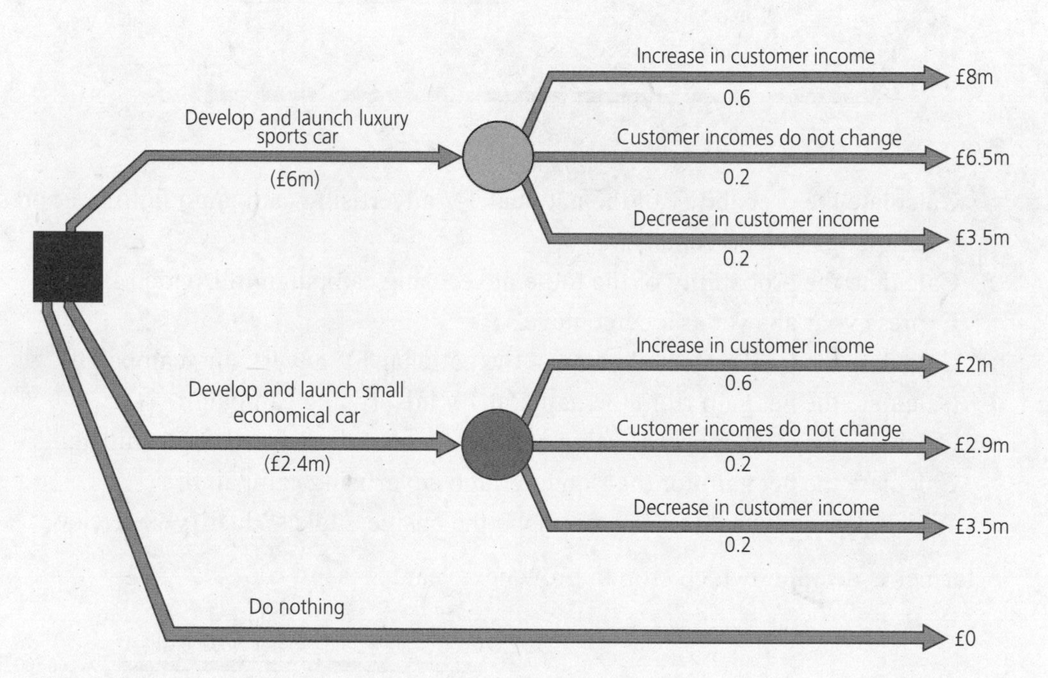

Figure 4.5

a Using the decision tree, calculate the net gain of launching a new sports car.

b Calculate the net gain of launching a smaller car.

c Based on your answers to parts **a** and **b**, which project would you choose?

A more senior marketing manager finds a different economic forecast. This one predicts a 40% chance of a rise in customer incomes, a 20% chance that customer incomes will remain the same and a 40% chance they will fall.

d Using these new figures, calculate the net gain of launching a new sports car.

e If the more senior marketing manager's economic forecast is correct, would you recommend the business launches a new sports car?

f Calculate the net gain of launching a new smaller car.

Critical path analysis

Note: this topic is for A-level candidates only.

Critical path analysis is a project management tool that allows businesses to identify:

- the prerequisites for carrying out a particular stage (tasks which must be completed before another task can begin)
- tasks which can be carried out simultaneously (at the same time)
- the lead time (how long the project will take)
- float times (how long non-critical activities can be delayed without delaying the overall project)
- critical activities (tasks which, if delayed, will delay the overall project). These form the critical path. These activities do not have any float time

Figure 4.6 is an example of a critical path analysis diagram (also known as a network diagram):

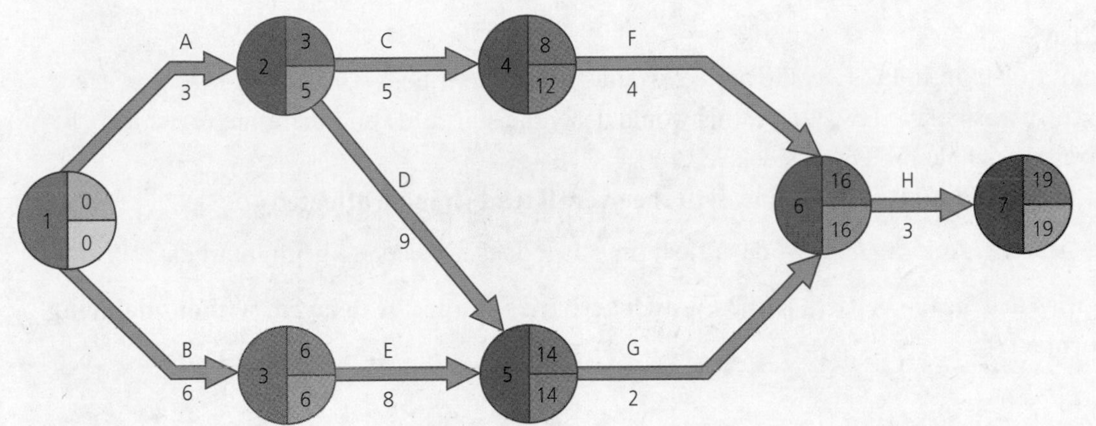

Figure 4.6

- Arrows represent activities, with the duration of the task below the arrow. Assume the duration is in days, in this case. It is also common for the duration to appear in weeks.
- The circles are nodes which represent the start of the next activity. The number on the left is the number of the node. The number has no particular significance but it makes communicating about the diagram easier.
- The number in the top right of each node is the earliest time that the next activity can start (earliest start time).
- The number in the bottom right of each node is the latest time the previous activity can finish, in order for the project still to run to time (latest finish time).
- Float time for an activity is calculated by using this formula:

  latest finish time – earliest start time – duration of the activity

A Worked example

Use Figure 4.6 to answer these questions.

i **Identify activities which can be carried out simultaneously.**

Activity A and activities B and E can be carried out at the same time. Activities C and F can be carried out at the same time as activity G.

ii **What is the prerequisite(s) to activity G?**

Before activity G can begin, activities D and E must be completed first.

iii **What is the lead time for the project?**

19 days

iv **What is the float time for activity A?**

latest finish time – earliest start time – duration of the activity

$5 - 0 - 3 = 2$ days

- If A did not finish until day five, the project would still run to time.
- However, if A was delayed by 3 days, this would delay the start of D and therefore G, which would mean the whole project was delayed.

v **If activity C is delayed by two days, how will the overall lead time be affected?**

It will not be affected. Activity C has 4 days float time as F does not need to begin until day 12.

vi **Which activities are on the critical path? (Which activities cannot be delayed, without delaying the overall project?)**

B, E, G and H

vii **Why is the earliest start time for G not 12 days? (A is 3 days and D is 9 days)**

- Though A and D are necessary for G to begin, activities B and E also need to be completed which take a total of 14 days ($6 + 8$).
- When calculating the earliest time an activity can start, add up the longest route. For example, H cannot begin until day 16. Even though A, C and F would be ready by then (after only 12 days), B, E and G take 16 days and H must wait until all these activities are complete.

B Guided question

Copy out the workings and complete the answers on a separate plece of paper.

1 **Figure 4.7 shows a critical path diagram. The duration of each activity is given in days.**

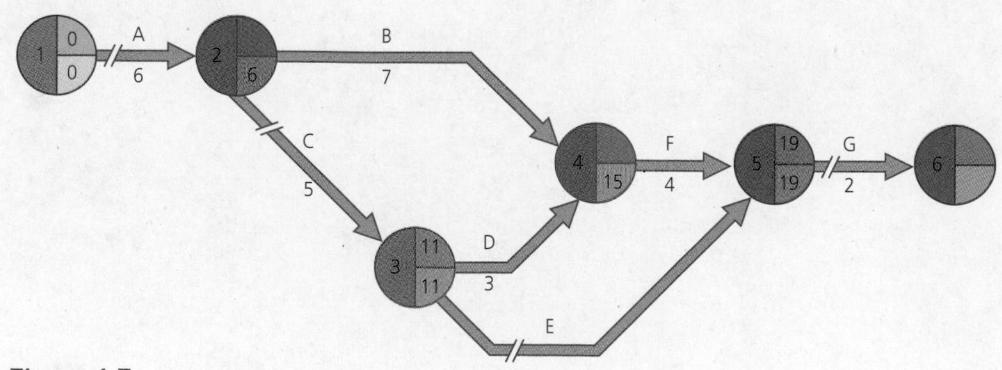

Figure 4.7

a What is the prerequisite for activity C?

- Activity C is shown by the arrow labelled C.
- Look at the arrows before activity C to see what must be done before activity C can begin.

b What is the earliest start time for activity B?

Look at the duration of the prerequisite for activity B.

c What is the earliest start time for activity F?

- Note that two sets of activities need to be complete to begin activity F: A+B and A+C+D
- Add up the duration for both sets of activities. Your answer will be the larger of the two numbers.
- Remember to give your answer in number of days.

d What is the duration of activity E?

- Note that activities A, C, E and G are on the critical path (shown by the // on the arrows).
- Look at the earliest start time for E and G.

e What is the lead time for this project?

- Lead time refers to the total time the project will take to complete and is found in the final node.
- Add up the duration of the activities on the critical path to find out the lead time.

C Practice questions

2 Using the critical path diagram in Figure 4.8, answer the following questions. The duration of activities is given in days.

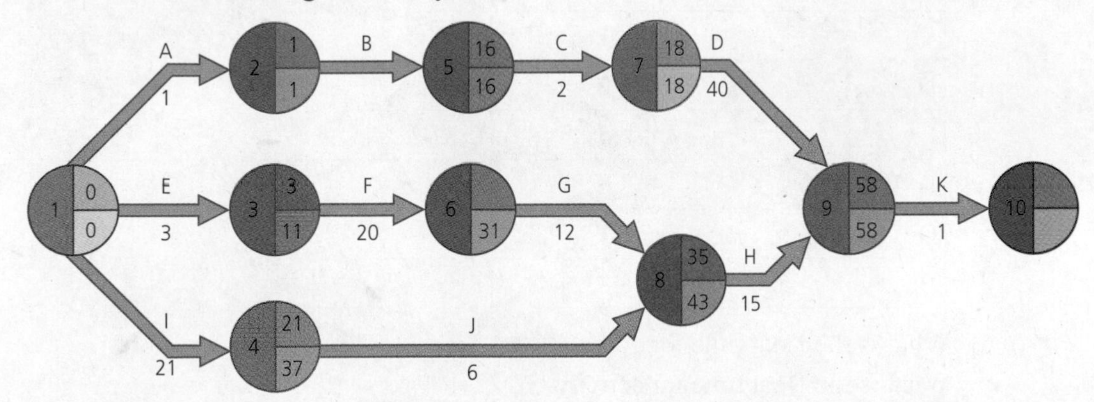

Figure 4.8

a What is the prerequisite for activity E?
b How long is activity B expected to take?
c What is the earliest time activity G can begin?
d What is the float time for activity I?
e What is the float time for activity D?
f Which activities are on the critical path?
g What is the lead time for the project?

3 Figure 4.9 shows a network diagram. The duration of each activity is given in days.

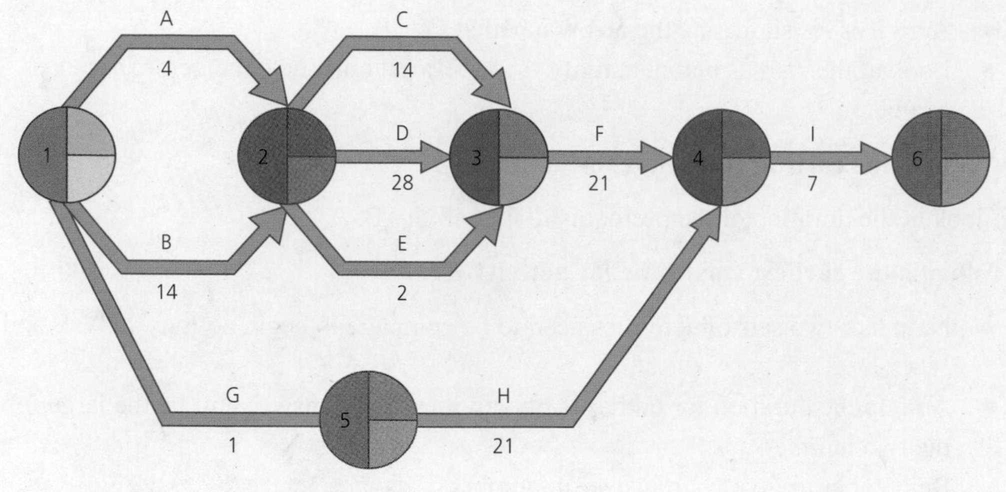

Figure 4.9

- **a** Copy the diagram and complete the earliest start times and latest finish times in the nodes.
- **b** Identify the activities on the critical path.
- **c** Calculate the float time for activity A.
- **d** Calculate the float time for activity C.
- **e** Calculate the float time for activity D.

4 a Draw a network diagram for the following project. Ensure all earliest start times and latest finish times are complete and the critical path is identified.

Table 4.5

Activity	Duration (weeks)	Prerequisite
A	5	–
B	10	–
C	1	A
D	2	B
E	5	C, D
F	6	E
G	1	E
H	8	F, G

- **b** What is the float time for activity C?
- **c** What is the float time for activity G?
- **d** The company has a policy of telling customers the lead time will be the actual lead time plus 13% to allow for a contingency. For this project, what would the company tell the customer the lead time is?

5 Human resources

Labour productivity

For many businesses, employees are their largest cost. Therefore, understandably, businesses want to get the most out of them as possible. Labour productivity is a common measure of workforce performance and refers to output per employee. This is easier to measure in some instances (for example, a factory worker) than others (a human resources manager).

Labour productivity is usually expressed as the number of units made (or sometimes sold) and is calculated by:

$$\frac{\text{total output}}{\text{number of employees}}$$

Labour productivity is essentially an average. You may want to revisit the section on Averages (page 6) before reading on.

A Worked example

A shift at a factory produced 80 cars. There were 30 employees at work at the time. Calculate the labour productivity.

$$\frac{80}{30} = 2.67 \text{ cars}$$

This means, on average, each employee made 2.67 cars. It is likely that some made more and others made less, or that they worked on a production line, each making a small part of all 80 cars.

Labour productivity has a knock-on effect for unit costs. The more employees make, usually the cheaper production costs are per unit. This increases a firm's profitability. Common ways of improving labour productivity include training, increasing motivation and installing more efficient machinery to help employees do their jobs. Businesses may measure labour costs per unit. These are calculated as follows:

$$\frac{\text{labour costs}}{\text{units of output}}$$

A Worked example

A call centre pays its employees £8.50 per hour. The human resources manager measured labour productivity (calls taken per employee per hour) yesterday and today. She found that yesterday an average of 15 calls were taken per hour per employee. Today 22 calls were taken per hour per employee. Calculate the labour cost per call taken yesterday and today.

Yesterday: $\frac{£8.50}{15} = 56.67\text{p}$

Today: $\frac{£8.50}{22} = 38.64\text{p}$

B Guided questions

Copy out the workings and complete the answers on a separate piece of paper.

1 A hospital is measuring the number of patients seen per consultant. After measuring the performance of the hospital's 54 consultants one day, it calculates the average number of patients seen per consultant to be 8.5 per hour. On average, how many patients were seen per hour at the hospital?

Multiply the number of patients seen per hour by the number of consultants.

2 A business is trying to increase productivity. Last month, employees worked 12 000 hours and manufactured 30 000 units in total. This month, employees worked overtime, working a total of 14 500 hours and manufacturing 8.75% more units than last month.

a Calculate how labour productivity, per hour, changed this month compared with last month.

Step 1: calculate labour productivity last month, per hour worked: $\frac{30\,000}{12\,000}$ = ________

Step 2: calculate the number of units produced this month by increasing last month's output (30 000 units) by 8.75%.

1% of 30 000 = 300

Step 3: calculate labour productivity this month. Remember to use this month's output and number of hours worked.

Step 4: compare labour productivity last month with this month's figure.

b What could explain this change?

C Practice questions

3 A hairdressing salon is measuring labour productivity for its stylists across three salons on one day. The results are shown in Table 5.1.

Table 5.1

	Porthmadog	Chwilog	Caenarfon
Number of stylists	5	4	9
Number of hair cuts	120	108	180
Daily wage per stylist	£71.10	£72.90	£64.80

a Calculate the labour productivity for:
 i Porthmadog
 ii Chwilog
 iii Caenarfon

b Calculate the labour cost per hair cut in:
 i Porthmadog
 ii Chwilog
 iii Caenarfon

c What is the relationship between the daily wage per stylist and labour productivity? What might explain this?

4 Table 5.2 is an index showing changes in labour productivity in the UK manufacturing industry. The manager of an LED lights factory compares changes in UK productivity as a whole, to changes in her own factory. Between 2015 and 2016, labour productivity at the LED lights manufacturer rose by 7%.

Table 5.2

Year	Labour productivity index
2014	100
2015	95
2016	102

Did labour productivity at the lights factory rise in line with changes in labour productivity for the UK as a whole, between 2015 and 2016?

5 A chemical plant in Bradford is trying to match the productivity of the firm's other plant in Holland. The productivity of the plants is shown in Table 5.3.

Table 5.3

	Bradford	Holland
Number of litres of final product produced per week	5 120	4 200
Number of employees	80	50

a Calculate Bradford's current labour productivity.
b Calculate labour productivity in Holland.
c How many litres of final product would Bradford need to produce to match Holland's labour productivity (without changing the number of employees)?
d How might the Bradford plant increase labour productivity?

6 A business is attempting to increase productivity. Currently employees are paid £9.00 per hour and labour productivity is 400 units per employee per day. Employees work eight hours a day. To try to boost motivation and productivity, management give all employees a 10% pay rise. As a consequence, labour productivity rises to 432 units per employee per day.
a Calculate the labour cost per unit before and after the pay rise.
b Has the decision to give employees a 10% pay rise been financially worthwhile for the business?

Labour turnover and retention

Employees leave employment at a firm for a variety of reasons. Perhaps an employee:

- gains a promotion by moving to another company
- is very unhappy in their work
- is able to get better rates of pay elsewhere
- is moving house and can no longer commute to work
- is retiring

Businesses usually want stability in their staffing with as few people leaving as possible. Though some 'fresh blood' brings in new ideas, generally recruiting large numbers of new staff is costly, disruptive and means expertise and skills in the workforce are lost.

Human resources departments often measure labour turnover and retention as a way of measuring how many people leave or stay employed at the business.

Labour turnover is the percentage of a business's employees leaving within one year (or other period of time such as a quarter). The formula is:

$$\frac{\text{number of employees leaving within a year}}{\text{average number of employees employed in the year}} \times 100$$

The average number employed within the time period can be calculated by:

$$\frac{\text{number of employees at the start of the year} + \text{number of employees at the end of the year}}{2}$$

A Worked example

A financially struggling business has 220 employees at the start of the year. Over the year, 24 employees leave. The business makes the decision not to replace them to save money. Calculate this firm's labour turnover.

Step 1: employees at the start of the year: 220

Step 2: employees at the end of the year: $220 - 24 = 196$

Step 3: average number of employees: $\frac{220 + 196}{2} = 208$

Step 4: labour turnover: $\frac{24}{208} \times 100 = 11.54\%$

Retention rates refer to the proportion of a firm's workforce that have worked for the business for a certain length of time, usually those who have worked at the business for a year or more. It is calculated by:

$$\frac{\text{number of employees who have worked at the business for a year or more}}{\text{number of employees at the end of the period}} \times 100$$

A Worked example

A company employs 12 000 people, 9560 of whom have been with the business for a year or more. Calculate the firm's retention rate.

$$\frac{9560}{12000} \times 100 = 79.67\%$$

B Guided questions

Copy out the workings and complete the answers on a separate piece of paper.

1 **A business has a staff retention rate (more than one year of service) of 60%. It has 435 employees. How many have been at the business for less than one year?**

Step 1: work out what percentage of 435 you are trying to find:

$100 - 60 = $ ______%

Step 2: find 1% of 435.

Step 3: multiply your answers to Step 1 and Step 2 to get the number of employees who have been at the business for less than a year.

2 A restaurant often employs students from the local college. These employees tend to leave after a few years to go to university or to other paid work and so the restaurant has high labour turnover. This year, labour turnover was 30% with 12 employees leaving. Calculate the average size of the workforce in the restaurant.

30% of the average workforce is 12 people.

1% of the average workforce is 12 divided by 30.

C Practice questions

3 Table 5.4 shows selected human resources data from a small chain of hotels.

Table 5.4

	Hotel A	Hotel B	Hotel C
Average number of employees	225	300	266
Number of employees leaving	25	35	21

a Calculate the labour turnover for:

i Hotel A

ii Hotel B

iii Hotel C

b Which hotel would you expect to have the lowest labour **productivity**? Why?

4 Details of a small firm's workforce are given in Table 5.5. Using this information, calculate the firm's retention rate.

Table 5.5

Employee	Length of service
Jenna	1 year
Joy	17 years
Javone	14 months
Jade	3 months
Julia	3 years
Jaabir	4 years
Jasper	1 week
Jenny	5 months
Jake	20 years

5 A call centre has, on average, 555 employees and a 60% labour turnover rate.

a Calculate how many employees left this year.

b Why might labour turnover be high in a call centre?

6 A business prides itself on its dedicated and motivated workforce. In one year, no employees leave and no employees join the team. The business has 60 employees. Calculate the firm's:

a labour turnover

b retention rate.

Absenteeism

Absenteeism is a measure of the proportion of working days lost due to employees not being in work. An employee may not attend work for a variety of reasons. They might be genuinely ill or have another authorised reason such a hospital appointment. Alternatively, employees may not attend because they dislike work. The more motivated a workforce is, the less absenteeism there should be. Human resources departments wish to avoid high levels of absenteeism as it places additional costs on the business, e.g. getting agency staff to cover staff shortages or placing stress on remaining employees by asking them to work additional hours.

Expressed as a percentage, absenteeism is measured by:

$$\frac{\text{number of days absent in the period}}{\text{total number of working days in the period}} \times 100$$

A Worked examples

a A business has ten employees, all of whom are full time. In a typical working (5 day) week, one employee has two days off work due to ill health. The others all attend. Calculate the absenteeism for the business in this week.

Step 1: number of days absent: 2

Step 2: total number of working days in the period: $5 \times 10 = 50$

Step 3: $\frac{2}{50} \times 100 = 4\%$

b There are 21 working days in a particular month. A business has 300 employees. The company loses 189 days from employee ill health that month. Calculate the rate of absenteeism.

$$\frac{189}{21 \times 300} \times 100 = 3\%$$

B Guided questions

Copy out the workings and complete the answers on a separate piece of paper.

1 In a month consisting of 23 working days, departments have the following number of days absent:

Table 5.6

Department	Number of employees	Number of days absent
Sales	15	50
Production	95	218
Finance	5	1

a Calculate the absenteeism for the sales department.

Step 1: calculate the maximum potential number of working days the sales department had:

$15 \times 23 = 345$ working days

Step 2: use the absenteeism formula to calculate the number of days absent as a percentage of total potential working days:

$$\frac{50}{345} \times 100 = 14.49\%$$

b Now repeat the process to find out the absenteeism for:

i the production department

ii the finance department.

Remember to first multiply the number of employees by 23 to identify the total number of possible working days.

2 A business is trying to reduce absenteeism by spending more on health and safety and employee welfare. Last year, before the new initiative, the average employee had 11 days off a year out of 222 days. This year the workforce of 78 employees had a total of 390 days absent from work.

Calculate the absenteeism before and after the new initiative.

Step 1: find last year's absenteeism.

$$\frac{11}{222} \times 100 = \_\_\_\_\_\_\%$$

Step 2: calculate the total number of working days this year by multiplying the number of employees by the number of working days in the year (222).

Step 3: use the absenteeism formula to calculate this year's absenteeism rate.

$$\frac{390}{} \times 100 = \_\_\_\_\_\_\%$$

C Practice questions

3 Out of 23 working days in a month, an employee attends work for 21 days. Calculate the absenteeism for this employee.

4 A manager studies rates of absenteeism in her branch for the last few months. The results are shown in Table 5.7.

Table 5.7

Month	Rate of absenteeism
October	14%
November	16%
December	

In December there were 21 working days and, on average, employees took 3 days off sick.

a Calculate the rate of absenteeism for December.

b Calculate the average absenteeism for the quarter.

5 A business has three employees on various contracts. The number of days absent per employee is outlined in Table 5.8, for one month consisting of 22 working days for a full-time employee. Calculate the absenteeism for this small workforce.

Table 5.8

Employee	Contract	Number of days absent
Alex	Part time. 0.5 contract meaning this employee works half a normal working week	2
Asha	Full time	1
David	Part time. 0.75 contract meaning this employee works three quarters of the hours of a full-time employee	0

6 A restaurant has five chefs working Tuesday to Saturday. This week the business has 8% absenteeism. Whenever a chef is absent for a day, the business calls an agency to send a temporary chef to cover their work. This costs the business £80 per day. This is 25% more than having their normal chef. Calculate the additional cost this business faces due to the absenteeism this week.

Exam-style questions

Section A: multiple-choice questions

Select one option only. One mark per question.

1 A market has a steady market growth. In 2014 the market was worth £3bn, growing to £3.3bn in 2015 and 3.63bn in 2016. If this annual rate of growth continues, what will the market's size be in 2018, to two decimal places? **(1)**

A 4.39bn **B** 4bn **C** 4.83bn **D** 3.99bn

2 Calculate this manager's profit variance using Table E.1. **(1)**

Table E.1

Income budget	£9 000	Expenditure budget	£2 000
Actual income	£10 000	Actual expenditure	£2 500

A £7000 favourable **B** £500 favourable
C £1500 favourable **D** £1000 adverse

3 A candyfloss stall has £100 fixed costs per day and £0.15 variable costs per unit of candyfloss sold. 300 units are sold for £1.50 each. Calculate the total contribution from these 300 units. **(1)**

A £74.07 **B** £305 **C** £405 **D** £300

4 Using the information in Table E.2, calculate this firm's operating profit margin. **(1)**

Table E.2

Number of units sold	180
Selling price	£4
Total cost of goods sold	£270
Operating expenses	£300
Tax paid	£30

A 62.5% **B** 16.67% **C** 20.83% **D** 4.8%

5 Here is a cash flow forecast for a business. Calculate this firm's forecasted closing balance for March. **(1)**

Table E.3

	January (£000s)	February (£000s)	March (£000s)
Total cash inflows	30	20	20
Total cash outflows	15	30	15
Net cash flow	15		
Opening balance	0		
Closing balance	15		

A £0 **B** £10 000 **C** £15 000 **D** £20 000

6 A business notices that when customer incomes fall by 10%, its sales rise by more than 10%. Which of the following is most likely to be the income elasticity figure for the firm's products? **(1)**

A −1 **B** −1.2 **C** −0.5 **D** +2

Note: question 7 is for Edexcel, OCR and WJEC/Eduqas candidates only.

7 A company has a workforce of 115 people who were all expected to be in work one week in late August. This working week was four days long due to a bank holiday Monday. In that week there were a total of 20 days lost due to employee ill health. Calculate the firm's rate of absenteeism that week to 2 decimal places. **(1)**

A 17.39% **B** 3.49% **C** 8.7% **D** 4.35%

8 On average, a firm's workforce comprises 4000 employees. During the year, 12% leave the company. Calculate how many employees left the firm that year. **(1)**

A 840 employees **B** 3520 employees
C 353 employees **D** 480 employees

9 A business sells 5000 units for £12 each. The firm has a gross profit margin of 20%. Calculate the total gross profit earned from the sale of these 5000 units. **(1)**

A £48 000 **B** £12 000 **C** £2.40 **D** £15 000

10 An entrepreneur is deciding whether to enter market A or market B or to enter neither. Calculate the net gain of entering market A using Figure E.1. **(1)**

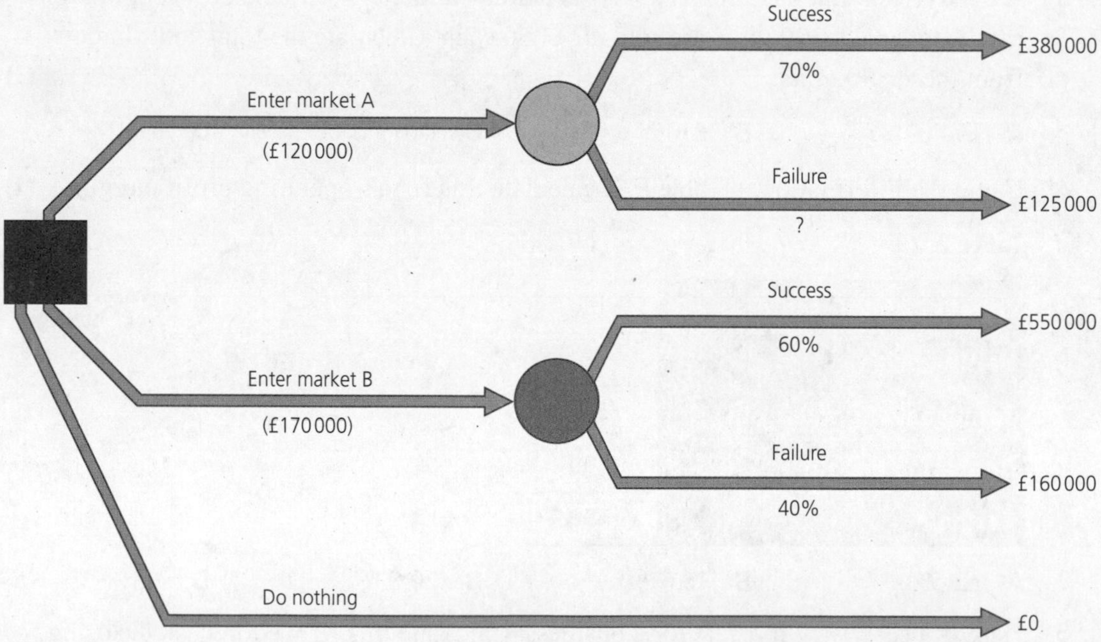

Figure E.1

A £183 500 **B** £303 500
C £30 230 000 **D** £196 000

Section B

You are advised to show your working and express answers in the appropriate units.

Extract A

Table E.4

	Last year	This year
Fixed costs	£266 800	£266 800
Average selling price	£4 250	£4 160
Variable costs per unit (cost of goods sold per unit)	£1 560	£1 590
Capacity	1 400	1 510
Number of units made/sold	1 380	1 400
Number of employees	15	16
Cotton price index	105	107

Silksby Ltd is a tailor on Savile Row (London) that makes high-end, bespoke suits for its clients. The tailor is well established in the area, operating from the same site for over 100 years. The tailor has an excellent reputation in London for customer service and traditional methods of manufacture, however, it must still work hard to compete with the 100 other tailors nearby. Table E.4 above gives some quantitative information about the business over the last two years.

Answer the following questions, using Extract A.

11 Calculate the tailor's margin of safety this year. **(4)**

12 Calculate the tailor's gross profit margin this year. **(5)**

13 Calculate the tailor's capacity utilisation:

a last year **(2)**

b this year **(1)**

14 What is the percentage change in costs per unit, over the time period shown? **(6)**

15 How has labour productivity changed this year, compared with last year? **(4)**

16 What is the price elasticity of demand for the tailor's suits? **(4)**

17 The business exports some suits to Japan. Last year the exchange rate was 1 Japanese Yen = £0.005. This year the exchange rate was 1 Japanese Yen = £0.008. In **Yen**, calculate how the price of one suit from the tailor changed this year, compared with last year. **(5)**

18 The formal menswear sector is worth £1.2bn. (Bn is short for 'billion'.) Calculate this tailor's share of that market this year. **(3)**

19 Using the cotton price index, calculate the increase in cotton prices, as a percentage. **(2)**

20 Explain two likely impacts of your answer to question 19 on the tailor. **(2)**

Extract B

Here is selected financial data for the tailor this year.

Table E.5

	This year
Fixed assets	£305 000
Inventory	£19 850
Receivables	£88 500
Cash	£110 000
Current liabilities	£20 000
Non-current liabilities	£1 338 750
Capital employed	£14 875 000

21 Using Extract B, calculate the tailor's current ratio. **(3)**

22 Using Extract B, calculate the tailor's gearing ratio. **(3)**

23 Using Extract A and Extract B, calculate the return on capital employed for this business, this year. **(4)**

Note: question 24 is for OCR and AQA candidates only.

24 Calculate the firm's receivables days. **(3)**

Extract C

Silksby Ltd is planning to expand in one of two ways. One option is to open a new store on the other side of London. A second option is to develop a website to allow customers from all over the world to buy suits online. This is something a few other tailors have started to do, with relative success.

Here are the predicted cash inflows and outflows for the two projects.

Table E.6

Year	New store		Website	
	Cash inflows	Cash outflows	Cash inflows	Cash outflows
0	£0	£250 000	£0	£80 000
1	£420 500	£310 000	£75 000	£2 500
2	£1 260 000	£790 000	£145 000	£2 500
3	£2 664 000	£1 330 000	£225 000	£2 500
4	£4 797 600	£2 050 000	£230 000	£2 500

Table E.7

Year	Discount rate @ 5%
0	1
1	0.952
2	0.907
3	0.864
4	0.823

25 Should the business open the new store or develop the website? Justify your view and use calculations to support your answer. **(20)**

Extract D

The business decides to open a new store, rather than develop the website. The operations director records the various stages involved in Table E.8 and in Figure E.2.

Table E.8

Activity		Duration (days)	Prerequisite
A	Find a suitable location	5	–
B	Negotiate terms and sign shop lease	2	A
C	Fit new floor	7	B
D	Decorate walls	5	C
E	Order sewing machines and other equipment	10	A
F	Hire employees	14	–
G	Send materials, e.g. business cards, to printers	10	B
H	Conduct marketing campaign	7	D, G, E, F

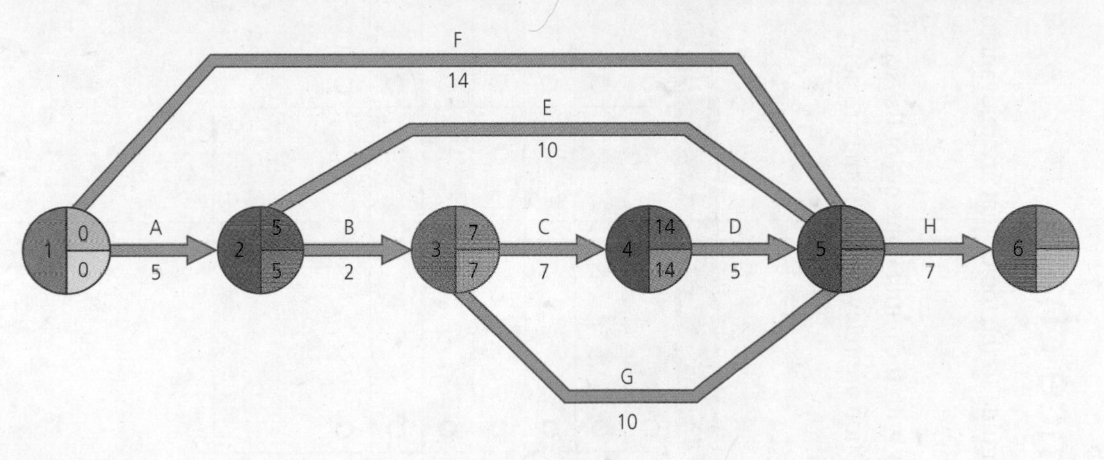

Figure E.2

26 State the order of the activities on the critical path. **(2)**

27 Calculate the lead time for the store opening. **(2)**

28 Calculate the float time for activity F. **(3)**

29 Calculate the float time for activity G. **(2)**

30 Assess two benefits to Silksby Ltd of creating a network diagram for this project. **(8)**

Appendix

Exam board cross-reference chart

The information in the following table is intended as a guide to the topics that are relevant to your awarding body. You should refer to your specification for full details of the topics you need to know.

- o A green circle indicates that you need to be familiar with the entire contents of the section.
- o An orange circle signifies that some parts of the section are relevant and some are not — further details are given when this is the case.
- o A red circle means the section is not relevant to you.

Unit	Topic	AQA	Edexcel	OCR	WJEC/Eduqas
1 Key mathematical skills	Averages	o	o	o	o
	Fractions	o	o	o	o
	Ratios	o	o	o	o
	Percentages	o	o	o	o
	Percentage change	o	o	o	o
	Interpreting graphs	o	o	o	o
	Interpreting index numbers	o	o	o	o

Unit	Topic	AQA	Edexcel	OCR	WJEC/Eduqas
2 Finance	Costs	O	O	O	O
	Revenue	O	O	O	O
	Profit	O	O *Edexcel uses a slightly simpler formula for profit for the year (net profit) of operating profit minus interest. This section is still relevant but you might face slightly less complicated questions in the exam.*	O	O *Your specification does not refer to operating profit. You should focus on gross and net profit.*
	Profit margins	O	O	O	O *Your specification does not refer to operating profit margin. You should focus on gross profit margin and net profit margin.*
	Contribution	O	O	O	O
	Break-even	O	O	O	O
	Budgets and variance analysis	O	O	O	O * *At AS level you need to know what is meant by a budget and its purpose. You are not required to understand and calculate variances until A-level.*
	Cash flow forecasting	O	O	O	O
	Exchange rates	O *	O	O * *At AS level you need an awareness of what an exchange rate is but you do not need to be able to calculate it or understand its impact.*	O *
	Liquidity ratios	O * *Your specification does not explicitly mention the acid test ratio so you may want to focus on the current ratio here.*	O	O	O *
	Return on capital employed and gearing ratios	O *	O *	O *Return on capital employed is part of the AS content but gearing is part of the A-level content. The elements of this section covering ROCE are relevant to AS level.*	O *
	Efficiency ratios	O *	O	O *	O
	Investment appraisal	O *	O *	O	O *

Unit	Topic	AQA	Edexcel	OCR	WJEC/Eduqas
3 Marketing	Market share, size and growth	O	O	O	O
	Price elasticity of demand	O *You need to be able to interpret price and income elasticity data and be able to analyse the impact of changes in price and income on revenue. You do not need to be able to calculate PED and YED figures.*	O	O	O * *At AS level you need an understanding of the concept of price and income elasticity but you are not required to calculate their numerical values until A-level.*
	Income elasticity of demand	O *You need to be able to interpret price and income elasticity data and be able to analyse the impact of changes in price and income on revenue. You do not need to be able to calculate PED and YED figures.*	O	O	O * *At AS level you need an understanding of the concept of price and income elasticity but you are not required to calculate their numerical values until A-level.*
	Forecasting	O *The extrapolation and correlation parts of this section are relevant to you. You do not need to study moving averages.*	O * *You need an awareness of sales forecasting at AS level but specific knowledge of moving averages and extrapolation is not required until A-level. Correlation is not explicitly mentioned in your specification.*	O*	O
4 Operations	Capacity and capacity utilisation	O	O	O *At AS level you need to know what is meant by capacity utilisation and why it is important to a business. You are not required to calculate capacity utilisation until A-level.*	O
	Decision trees	O	O*	O*	O*
	Critical path analysis	O*	O*	O*	O*
5 Human resources	Labour productivity	O	O	O	O
	Labour turnover and retention	O	O*	O *Your specification requires knowledge of labour turnover but does not explicitly mention retention. You may want to focus on the labour turnover parts of this section.*	O *Your specification requires knowledge of labour turnover but does not explicitly mention retention. You may want to focus on the labour turnover parts of this section.*
	Absenteeism	O	O*	O	O

*concept required for A-level Year 2 only